Incremental Polarization

A Unified Spatial Theory of Legislative Elections, Parties, and Roll Call Voting

JUSTIN BUCHLER

OXFORD
UNIVERSITY PRESS

OXFORD
UNIVERSITY PRESS

Oxford University Press is a department of the University of Oxford. It furthers the University's objective of excellence in research, scholarship, and education by publishing worldwide. Oxford is a registered trade mark of Oxford University Press in the UK and certain other countries.

Published in the United States of America by Oxford University Press
198 Madison Avenue, New York, NY 10016, United States of America.

Library of Congress Cataloging-in-Publication Data
Names: Buchler, Justin, author.
Title: Incremental polarization : a unified spatial theory of
legislative elections, parties and roll call voting / Justin Buchler.
Description: New York, NY : Oxford University Press, [2018] |
Includes bibliographical references.
Identifiers: LCCN 2017045024 (print) | LCCN 2018000700 (ebook) |
ISBN 9780190865603 (updf) | ISBN 9780190865610 (epub) |
ISBN 9780190865580 (hardcover) | ISBN 9780190865597 (pbk.)
Subjects: LCSH: United States. Congress—Elections. | United States.
Congress—Voting. | Political parties—United States. |
Polarization (Social sciences)—Political aspects—United States. |
Spatial analysis (Statistics)
Classification: LCC JK1975 (ebook) |
LCC JK1975 .B83 2018 (print) | DDC 328.73/0769—dc23
LC record available at https://lccn.loc.gov/2017045024

Incremental Polarization

CONTENTS

Incremental Polarization

1

Introduction

Ideological polarization in Congress has reached historic levels (Figure 1.1). Ironically, as that polarization has developed, scholars have come to rely increasingly on spatial theory to explain elections in general, and legislative elections in particular. The problem is that the central tendency of an election governed by spatial theory, at least in a unidimensional context, is for candidates to converge to the location of the median voter,[1] so spatial theory's prominence has increased as its ability to predict outcomes has become more suspect. Why do legislative candidates take ideologically extreme positions? How do they win given their extremism? And perhaps most difficult for spatial theorists, where do political parties fit in the picture? Spatial theorists have attempted to resolve each of these basic questions in a piecemeal fashion, thereby leaving critical gaps in our understanding of legislative polarization. The objective of this book is to present a unified spatial theory of legislative elections, parties, and roll call voting. Only by linking spatial models of elections to spatial models of roll call voting in a unified model can we provide a cohesive explanation for the strategic, electoral, and procedural roots of polarization, and the critical role that parties play in the process.

The Problem of Polarization

When Anthony Downs published his treatise in 1957 on applications of a Hotelling-style spatial model to electoral politics, he did so in an era of apparent centrism that suggested the very pressure implied by Hotelling's method. There were two problems with that, which would become more evident as spatial theory advanced and as historical trends emerged. First, the ideological overlap between the parties during the mid-20th century was a historical anomaly. If anything, polarization is actually the norm. To propose a model in the mid-20th century to explain the politics of the era clearly misses something about its deviation from historical patterns. The second problem is that the policy space is not truly unidimensional, and as Poole and Rosenthal (1997) demonstrated,

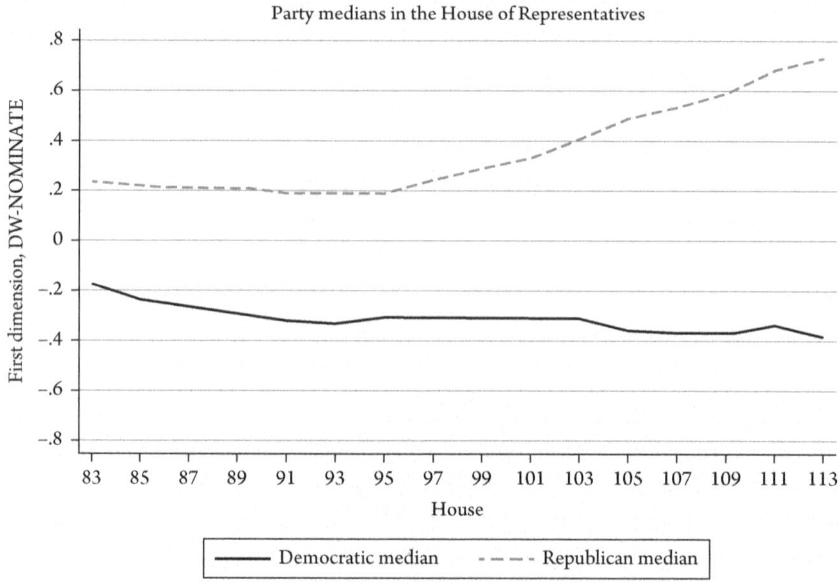

Figure 1.1 Party medians in the House of Representatives

the mid-20th century was characterized not merely by parties converging to the center along the first dimension of NOMINATE (ostensibly liberalism-conservatism), but by intraparty divisions along the second dimension, which involved the complicated politics of race, region, and party in that era. The process by which the parties have moved apart along the first dimension is also the process of policy disagreements collapsing onto the first dimension as the liberal-conservative divide explains a greater proportion of roll call votes in Congress. As Noel (2014) argues, the definitions of liberalism and conservatism change over time as coalition merchants and policy demanders bind previously uncorrelated issues together to create what we now know as liberalism and conservatism, and the current liberal-conservative spectrum.

While the development of legislative polarization is not simply the process of Democrats moving left and Republicans moving right, though, the left-right divide has become the primary focus of legislative scholarship, and understandably so. While there is broad agreement on the concept of the liberal-conservative spectrum in American politics, the meanings of the other dimensions, such as they remain, are less clear. The second dimension of NOMINATE, for example, has involved the complicated politics of race and region, which have aligned more along the first dimension over time anyway, but its meaning is less clear in the 21st century. In more practical terms, spatial models are easily solved in one dimension, but adding a second dimension introduces the problem of the collective intransitivity of preferences and a form of "chaos" (McKelvey 1976;

Schofield 1978) that leaves equilibria as either structurally induced (Shepsle & Weingast 1981) or nonexistent.

Nevertheless, the liberal-conservative spectrum explains an increasing proportion of roll call votes over time, and along that spectrum the parties are as far apart as they have ever been. And since legislators who vote as extremists along that spectrum receive lower vote shares than their "centrist" colleagues, which is as the unidimensional models predict, the trend towards polarization in Congress remains a challenge for spatial theory.

Incrementalism and the Development of Polarization

Legislative polarization must be understood incrementally because it develops incrementally. Our primary measures of ideology in Congress are built around roll call voting patterns. When we say that Congress is ideologically polarized, what we really mean is that congressional *voting scores* are polarized in a way that suggests ideological divergence. Whether those patterns, measured through NOMINATE, indicate true ideology in a utilitarian sense, or partisanship, as Lee (2009) may assert, the problem remains that "extreme" scores are associated with lower vote shares, indicating electoral pressure toward centrism, or some facsimile of it. So why don't legislators vote in a way that at least mimics centrism?

Note how this question reframes spatial positioning. Winning legislative candidates do not make single-stage declarations of policy that provide their constituents with policies equivalent to their locations in the policy space as soon as the election is over, as traditional spatial models imply. Instead, legislators adopt positions in the policy space, polarized or otherwise, through the incremental process of casting roll call votes. Thus, understanding the development and persistence of polarization in Congress requires understanding why legislators incrementally build ideologically extreme voting records, and why policy-motivated voters respond to these voting records in such surprising ways. Once we understand that process, it will become clear that only through party leadership in Congress can it be rational for ideologically extreme legislators to adopt sincerely extreme positions in the policy space, and only through the roll call voting process can we understand electoral responses to the incremental adoption of extreme positions.

The necessity for party leadership comes from the fact that sincere voting in Congress can be a collective action problem. Consider the House of Representatives, where polarization has become especially pronounced, and where the book will direct its primary focus. Suppose that members of a party caucus in the House have homogeneous, noncentrist ideological preferences over policy, but that they represent districts with a wide range of preferences. Some represent districts whose constituents share the noncentrist preferences of the legislative caucus, but others represent districts with more moderate preferences. In the absence of party leadership, such a caucus will not vote cohesively on roll calls, and as a result it will fail to achieve its policy objectives. The obstacle to cohesion is the fact that most roll call votes are not decided by a single-vote margin. Unless the rest of the House of Representatives is divided 217 for and 217 against, any member of Congress representing a moderate district should vote in such a way as to please her moderate constituency regardless of how extreme her policy preferences are because to do otherwise would entail paying an electoral penalty for no policy benefit since the policy outcome will not depend on the legislator in the question's vote. As a demonstration, suppose that all members of a 235-seat Democratic House caucus would personally prefer for a bill to pass. However, the bill is a very liberal one, so only actors with relatively liberal ideal points in the policy space will prefer the bill to the status quo. Suppose that 75 members of this House caucus would optimize their chances of re-election by voting against the bill because their constituents are not sufficiently liberal to prefer the bill to the status quo. In the absence of party leadership, the bill should only receive 160 votes, which is not enough to pass. The reason is that none of the 75 cross-pressured legislators has any incentive to deviate from this arrangement because if only 160 legislators are voting for the bill, becoming the 161st supporter would require casting an electorally suboptimal roll call vote without actually yielding any policy benefit. The Nash equilibrium solution is for all cross-pressured legislators to vote against the bill, causing the bill to fail even though every member of the majority caucus would personally prefer for the bill to pass.[2] Thus, for an ideologically homogeneous but electorally diverse legislative caucus, sincere voting is a collective action problem, and without someone to solve that collective action problem, the caucus will fail to achieve any noncentrist policy objectives its members may have.

The solution is for parties to function as "preference-preserving institutions." If caucus members attach sufficient importance to policy outcomes, they will empower party leaders to solve the collective action problem of sincere voting through a disciplinary system that encourages sincere voting, either by punishing legislators who cast insincere roll call votes,[3] or by rewarding legislators who cast sincere but electorally dangerous roll call votes. For various reasons, the model will focus on punishments, although the underlying principles of the model would work equally well either way. In the hypothetical example above, that means empowering party leaders to threaten the 75 cross-pressured legislators (or at least 58 of them) with modest punishments if they attempt to "free-ride" by voting insincerely against the bill. That party discipline, despite the modesty of the threats, will lead to passage of a bill that caucus members personally support, but that many of their constituencies oppose. Thus, party discipline will allow the caucus to achieve policy goals that it could not otherwise achieve, and that even the 75 (or 58) subjects of the disciplinary measures want to achieve. However, the result of that discipline is that caucus members *incrementally* adopt somewhat more sincere, ideologically extreme, and electorally suboptimal locations in the policy space. Yet the overall structure of the legislative agenda will create electoral paradoxes that reduce the damage of extremism, allowing most to win re-election anyway with only somewhat lower vote shares. And, since the caucus is policy-motivated, that will be considered an acceptable trade-off.

Since the mid-20th century there has been an increase in ideological homogeneity within the two party caucuses, but contrary to conventional wisdom, we have observed little increase in the electoral homogeneity of the caucuses. As that state of affairs has developed, sincere voting has increasingly become a collective action problem, giving policy-motivated legislators more incentive to delegate power to party leaders to solve that problem. The trend toward polarization in congressional voting scores, then, has resulted from the development of the conditions under which policy-motivated, cross-pressured legislators need party leadership to solve the collective action problem of sincere voting, along with legislative agendas that mitigate the electoral damage of the extreme positions that legislators incrementally adopt in the process.

This pattern began to change in the Republican caucus after the 2010 election. Republican leaders wound up in conflict, not with their moderate wing (such as it is), but with their most conservative wing. However, the reason was not a fundamental change in the nature of incumbent positioning, nor the purpose of party leadership. Moreover, while popular commentary on post-2010 divisions within the Republican Party focused on caucus members' newfound fear of primary challenges, that too is insufficient to explain the change in the nature of party leadership after 2010. Rather, a combination of extreme polarization and divided government eliminated incentives to work on normal legislation between 2010 and 2016, shifting the legislative agenda to items that uncovered previously hidden ideological heterogeneity while creating a different set of collective action problems for party leadership to solve. Those developments in Congress, then, merely reinforce the need to connect spatial positioning models of congressional elections to the structure of the legislative agenda and the roll call voting process.

Understanding why legislators adopt extreme positions and how they win re-election given their extremism requires a unified spatial model of legislative elections, parties, and roll call voting that explains the incremental process by which polarization develops through roll call votes, the effect of party leadership on those votes, and the paradoxical ways that rational voters will respond to legislative voting records.

Of course, spatial theorists have been grappling with these issues for years. The result has been a patchwork of mathematical tricks, divided between models of elections and models of legislative procedure that are individually useful, but collectively unable to provide a cohesive explanation for polarization at the institutional level precisely because they fail to connect the process by which ideologically extreme legislators get elected to the legislative process by which they manifest their extremism once they are elected. Spatial theory cannot come to grips with these puzzles without a unified model linking the electoral process to the roll call voting process.

The Disjointed State of Modern Spatial Theory

The persistence of the division between spatial models of elections and spatial models of roll call voting may actually be more puzzling than polarization itself. First, consider conventional spatial models of elections. In these models, each candidate's strategically selected platform is represented by a single point in an N-dimensional policy space (usually, as in the case here, a single dimension), which represents the policies that they would pursue if elected. Each voter has a utility function defined for all locations in that policy space, and the utility that a voter will receive from any given candidate's victory is the utility that the voter derives from that candidate's location in the policy space. Mathematically, then, most spatial theories of elections are based on the premise that the winning candidate provides all voters with a policy equivalent to her location in the policy space with a single action once the election ends.[4]

The problem with this premise is that it disregards the substantive process by which winning legislative candidates actually carry out their campaign platforms. First and foremost, winning legislative candidates cast roll call votes, so in order for a winning candidate to carry out her platform, that is, her location in the policy space, she must cast a series of roll call votes in a manner that is consistent with that location.

Ironically, of course, the other primary use of spatial theory in the legislative context is to study exactly that process. We can characterize the range of policy options with an N-dimensional policy space. Each roll call vote can be characterized by a status quo point within that policy space, describing the current state of policy, and an alternative point in the policy space, describing what policy would be if the bill in question passed. Each legislator's position can also be characterized by a point in the policy space. When the location of the status quo is closer to a legislator's position, that legislator will vote against the bill, and when the location of the alternative is closer to a legislator's position, the legislator will vote for the bill. Thus spatial models of legislative behavior treat a roll call vote in the same manner that spatial theories of elections treat candidate elections.

In fact, spatial theories form the basis of how we *measure* legislators' locations in the policy space. The NOMINATE procedure developed by Poole and Rosenthal estimates legislators' locations in a policy space by calculating the locations that best explain the votes that each legislator casts. These calculations are based on the *assumption* that each roll call vote is a choice between a status quo and alternative point within the N-dimensional space, and that each legislator will vote for the option closest to her location in that policy space. The near-circularity of that reasoning forms the basis of critiques of NOMINATE.

Models of legislative agenda control operate on the same principle. Actors with positive agenda-setting power will force votes on any bill that would move policy toward their ideal points and pass, while actors with negative agenda-setting power will block consideration of any bill that would move policy away from their ideal points and pass.

The problem with these spatial models is that they ignore the process by which legislators select their locations entirely. They simply examine roll call voting decisions and other aspects of the legislative process while taking legislators' ideological positions, extreme or not, as fixed and exogenous despite the fact that these positions may actually be the result of the interaction between electoral considerations and the very legislative processes that these models purport to study. We can measure aggregate roll call voting patterns based on the assumption that the process is a simple spatial calculation, and those patterns have a great deal of predictive power, but omitting elections from the study of roll call voting makes it impossible for us to understand the actual origins of those patterns. It becomes pure description, and we cannot explain the rise of polarization in these patterns simply by describing the patterns.

Thus, each of the two common applications of spatial theory to the study of Congress has a critical gap. Spatial models of elections ignore the centrality of roll call voting to the positioning process, and spatial theories of roll call voting ignore the complex interactions between legislators' personal preferences, electoral considerations, and party pressure that combine to form a legislator's observed location in the policy space. Serendipitously, then, the critical gap in each application of spatial theory is precisely what the other application can help us to understand. Thus by unifying these two strands of spatial modeling into a cohesive theory of legislative elections and roll call voting, we can fill the gap in each of the component models while explaining both why candidates adopt extreme positions and how they win.

The Underlying Principles of a Unified Spatial Theory of Congress

Existing spatial models of elections and existing spatial models of legislative roll call voting are structurally compatible and conceptually complementary. They can be combined into a unified model once we understand one basic observation: the candidate who wins a legislative election does not implement her platform in a single action once the election is over, as conventional Downsian models assume. Instead, she implements her platform incrementally, by casting a sequence of roll call votes. The methodological proposition of this book, though, is that we should turn that observation around, and look at the relationship between roll call voting and electoral positioning from the perspective of incumbent members of Congress rather than from the perspective of nonincumbent candidates. For incumbents, the roll call voting process is not one that *follows from* the policy location a member adopts. For sitting members of Congress, the roll call voting process *is* the process by which they adopt their electoral locations in the policy space.

This observation is the central organizing principle for the unified model, and it suggests a new approach to both spatial models of elections and spatial models of legislative procedure. Traditional spatial models of elections place the election before the legislative session, and then treat the legislative session as a black box. Traditional spatial models of roll call voting in Congress unpack that black box, but largely disregard elections as though legislators' locations were exogenous rather than determined by that legislative process in anticipation of electoral implications. This book will link spatial models of elections to spatial models of the legislative process, but in reversed order. Rather than an election, the first stage of this model will be a legislative session. The session will be held under the assumption that the majority party leader has procedural control and disciplinary tools, making the model better suited to explaining the House of Representatives than the Senate, but many of the mathematical results will have important implications for the Senate as well. The first stage of the legislative session will be an election within a majority party caucus for a leadership position. Leadership candidates will formulate platforms consisting of disciplinary systems. The winning leadership candidate will then apply her disciplinary system to a legislative agenda that permits majority-sponsored bills but prohibits minority-sponsored bills. The chamber will then vote, sequentially, on each item on the legislative agenda. Each member of Congress will have to weigh her personal policy preferences, electoral incentives, and party discipline threats on each vote as she builds her voting record, and hence incrementally establishes her location in the policy space. Finally, the elections will occur in which voters in each district evaluate their incumbents based not on their locations in the policy space, but on the bundles of roll call votes used to construct those locations, thereby linking spatial theories of roll call voting to spatial theories of elections in a unified model.

Reversing the order of events will be a dramatic departure from standard spatial methodology, but it follows logically from a relatively forgotten observation by Downs (1957) that a party's history constrains its ability to offer a new platform. Reversing the order of events merely formalizes this observation, but it allows us to explain why incumbents adopt extreme locations and why they win. As we shall see, the problem has been that the conventional structure of spatial models has hidden some critical factors in the development of polarization. By placing elections before the legislative session, conventional models have drawn attention to the strategic interaction between competing candidates operating within the same electoral context. Positioning decisions, then, are the result of campaign dynamics, and polarization must be reconciled with the logic of campaigns. While this strategic interaction is both real and important, the consequence of focusing on it has been to draw our attention *away* from another form of strategic interaction that is at least as important—the strategic interaction between sitting legislators operating across different electoral contexts. The legislative process and the nature of party organization in Congress are determined by that strategic interaction. Moreover, since the positioning process *is* the roll call voting process for incumbent members of Congress, electoral positioning is not merely a function of campaign dynamics. It is also function of legislative dynamics, and by placing the legislative session before the election, we can explain both the development of extreme spatial positioning decisions and electoral responses to those decisions in terms of intraparty legislative dynamics.

Why does conventional spatial theory place the election first? Tradition. Every election except 1788 was preceded by a full Congress, so unless our intent is to model the complete history of Congress with a single model, the question of whether to place the legislative session first or the election first is a chicken-and-egg question. The argument of this book is not that we must always place the legislative session first. Rather, adherence to the election-first structure has blinded spatial theorists to the distinctive and complementary insights that come from placing the legislative session first. This book aims to fill that gap, with the goal of allowing different models to complement each other rather than compete with each other.

Consider the electoral success of extremists. Conventional spatial theory poses the question as follows: Why do voters elect candidates with extreme policy platforms? Reversing the order of events changes the nature of the question. Instead, the book will ask, why do voters *re*-elect incumbents who have incrementally adopted extreme positions through their roll call voting patterns, even when faced with more centrist challengers? That is a subtly different question. The answer that this book will propose is that the legislative agenda can create what shall be called a "proximity paradox." As we shall see, even a hypersophisticated voter can paradoxically prefer the bundle of roll call votes offered by a distant incumbent to the bundle offered by a more proximate challenger, allowing extreme incumbents to defeat centrist challengers. Incumbents will receive modestly lower vote shares for their extremism, but still frequently win given these paradoxes

We will also see why legislators are willing to adopt electorally suboptimal positions in the policy space *in anticipation* of these diminished vote shares, and why parties are central to that phenomenon, thereby resolving the remaining puzzles. In fact, placing the legislative session first will show how strategic legislators set up the legislative process, through party organization, in order to *make* such positioning rational. When a legislative caucus is ideologically homogeneous, electorally diverse, and policy motivated, its members will empower party leaders to impose punishments on legislators who vote insincerely for electoral purposes. Doing so will allow the caucus to achieve its noncentrist policy goals. The consequence, though, will be that caucus members incrementally adopt ideologically extreme and electorally suboptimal positions in the policy space. However, the structure of the legislative agenda will reduce the electoral costs of that extremism by creating the mathematical paradoxes mentioned above. Thus, the caucus will pay only modest electoral costs for their extremism, allowing most to win *re*-election even though their positions in the policy space are electorally suboptimal. Moreover, the value they ascribe to policy outcomes makes it rational for them to sacrifice some electoral security in order to achieve those policy victories.

Thus, by linking spatial models of elections to spatial models of roll call voting, and by reversing the order of conventional positioning models, we can explain why legislators adopt extreme positions, how they win re-election given their extremism, and why parties are necessary for the development of polarization in Congress.

Parties and Spatial Theory

While the most dramatic departure from conventional spatial methodology will be placing the legislative session before the election, the other major departure will be carving out a central role for political parties in Congress, thereby addressing not only questions of polarization, but questions about the nature of party influence in Congress. How much influence do parties have on political outcomes in Congress, and how much of that influence comes from pressure on substantive roll call votes rather than mere procedural control?

Party-based theories of Congress are dominated by the Cartel model (Cox & McCubbins 1993, 2005) and Conditional Party Government (Aldrich 1995; Aldrich & Rohde 2000; Rohde 1991), and the unified model here will operate primarily within the CPG tradition. In the CPG model, members of a party caucus delegate power to party leaders to manage the legislative process when a set of conditions are met, the most important and controversial of which is that the caucus is internally ideologically homogeneous. Intuitively, the condition makes sense because it would be illogical for an ideologically diverse party to permit centralized authority. However, critics of the theory, particularly those who adhere to the party-less Pivot Point model (Krehbiel 1998), argue that if a majority caucus is ideologically homogeneous, then its members will vote cohesively on roll call votes and achieve the same policy outcomes regardless of whether or not the party applies the kind of procedural control that skeptics say doesn't exist anyway. If so, then CPG borders on tautology.

Moreover, both the Cartel model and CPG tend to focus on the means by which parties structure the legislative process rather than the pressure they apply to legislators on votes, and even when it comes to their influence on roll call votes, party-based models tend to focus on the influence that parties apply on procedural rather than substantive votes (see, for example, Cox & McCubbins 1993, 1994, 1997, 2002, 2005; Jenkins et al. 2005; Sinclair 2011; Theriault 2008; Van Houweling 2012).

In order to explain the role that parties play in legislative polarization, the unified spatial model will rethink the nature of party influence on roll call votes, and in the process, absolve CPG of the tautology charge. Discussion of party influence on roll call voting is generally tainted by the implicit assumption, made more dangerous by going unstated, that "influence" means pressuring legislators to cast roll call votes against their sincere, personal policy preferences. However, there are two different types of influence that parties can apply: "preference-deviating influence" and "preference-preserving influence." When parties apply preference-deviating influence, they pressure legislators to cast votes that conflict with their sincere, personal policy preferences. However, when parties apply preference-preserving influence, they pressure legislators to *disregard* the *electoral* pressure to vote insincerely. The function of preference-preserving influence is to cancel out electoral pressure. Thus, the result of preference-preserving influence is that legislators vote according to their sincere policy preferences when they wouldn't otherwise do so because they face other pressures.

Conceiving of parties as "preference-preserving institutions" will allow the derivation of a new set of conditions for party government that absolve CPG of the tautology charge. The key will be a distinction between ideological homogeneity and electoral homogeneity. A caucus is ideologically homogeneous when its members have similar personal preferences over policy, and a caucus is electorally homogeneous when its members represent politically similar districts. Under the unified model, a caucus will delegate power to the party leader to apply preference-preserving influence when the caucus is ideologically homogeneous, electorally *hetero*geneous, and policy motivated. The combination of ideological *homo*geneity and electoral *hetero*geneity turns "sincere" roll call voting into a collective action problem, as discussed earlier, and the strength of legislators' policy preferences gives them incentives to delegate power to party leadership to solve that problem. The key to absolving CPG of the tautology charge is that if a caucus is ideologically homogeneous but electorally heterogeneous, then in the *absence* of party leaders applying preference-preserving influence, they will *not* vote cohesively on roll call votes because sincere voting is a collective action problem. Thus, empowering party leaders when these conditions are met leads to different policy outcomes than failure to empower party leaders. Moreover, this suggests that if legislators vote *sincerely based on their policy preferences* with no additional predictive power for party labels, that is actually a manifestation of CPG rather than a demonstration of party weakness because in the absence of party influence, roll call votes would not be sincere.

Additionally, this model of party influence will resolve a central question about party strength in the US Congress. Parties are comparatively weak in the US Congress because unlike in parliaments structured around proportional representation (PR) electoral rules or the Westminster system, party leaders do not select nominees. Instead, party leaders are *selected by* independently nominated legislators. With the occasional exception, the House of Representatives consists of 435 members who independently won a major party primary and then independently won a general election against the opposing party's nominee, and who can win re-election through the same process even if their own party leaders don't want them in Congress. The Speaker of the House cannot remove intransigent caucus members from Congress altogether by removing their names from the ballot. Rather, the Speaker, who is elected by the majority caucus, must worry about having the gavel taken away by her caucus. Thus, when Speaker Cannon alienated enough members of his own caucus, he was stripped of his powers as Speaker in 1910, and when Speaker Boehner lost the support of his conservative wing, he too was forced to step down in 2015.

In contrast to legislative leaders in other countries, then, leaders in the US House face two limitations on their capacity to impose discipline. First, they have comparatively weak tools available to them, and, second, use of those tools risks alienating their caucus, whose support they need to maintain their positions. Building a unified spatial model of legislative politics around party discipline, then, has unique challenges with respect to the US House of Representatives.

In recognition of these challenges, the unified model will stack the deck *against* the rationality of party discipline by focusing on disciplinary systems built around punishments rather than rewards. Thus, by showing not only that party leaders can achieve dramatic effects while making only modest threats, but that rational legislators will *demand* such discipline, the argument will be stronger. In fact, when the reconceived conditions for party government are met, party leaders will not risk a 1910-style revolt for imposing party discipline—they will risk such a revolt if they *fail* to do so, even with the unique constraints that US legislative leaders have on their capacity to impose discipline. The 2015 revolt against Speaker Boehner, though, will show how even a subtle change in the legislative agenda will upset this delicate balance because of the intrinsic weakness of US legislative leaders.

Differences between Elections for Different Offices

This book introduces a spatial model that is uniquely applicable to legislative politics because it links the electoral process to the process of casting roll call votes in a legislature. Moreover, while many of the observations will be equally applicable to the Senate, the model is intended primarily to explain elections and behavior in the House of Representatives. Adherents to spatial theory as a general model of politics are likely to be frustrated by that aspect of the model. However, there is an important point here for the very nature of electoral theory, spatial or otherwise. The quest for a generalized theoretical model of electoral politics, applicable to any office, is fundamentally misguided. The central principle behind the model here is that the primary method by which legislators actually position themselves within the policy space is the process of casting roll call votes, so a theoretical model of a legislative election must reflect that. More broadly, then, a theoretical model of an election must consider the specific actions that will be taken by the winner, or similarly, the actions that *were* taken in the past by the incumbent. Executives do not cast roll call votes. Presidents and governors can sign or veto bills that reach their desks, but they have different responsibilities that directly influence policy through means such as executive orders. Thus a model of executive elections, spatial or otherwise, must characterize such actions in a realistic and theoretically coherent manner. Spatial theory has become popular in part because it is supposed to be generalizable to any type of election. However, if the predictions of a spatial model change dramatically by modeling the actual behavior of the officials in question rather than using an abstraction that doesn't reflect the behavior of any specific elected official, then there can be no generalizable theoretical model of an election. A theoretical model of an election cannot hope to capture the important aspects of the electoral process without taking into consideration the means by which the winner of the election will actually implement her platform, or by which the incumbent *did*, and since the means of doing so depend on the office in question, so must the model. The model in this book is uniquely applicable to legislative elections. No realistic model can be more generalizable than that, however frustrating that may be to those in search of a universal model of electoral politics.

Layout of the Book

This book is structured around resolving three basic puzzles about the relationship between legislative polarization and spatial theory. Why do members of Congress adopt ideologically extreme, electorally suboptimal positions in the policy space, how do they win re-election despite their extremism, and how do parties affect the process? The model will show that an ideologically homogeneous, electorally diverse, policy-motivated caucus will empower party leaders to solve the collective action problem of sincere voting in the legislature. As a result, legislators will incrementally adopt ideologically extreme, electorally suboptimal locations in the policy space. Consequently, incumbents will receive lower vote shares than they would had they not empowered parties to act in this manner, but they will still win re-election most of the time anyway because of a paradox in which voters prefer the bundles of roll call votes associated with distant locations over the bundles associated with more proximate locations. They simply win re-election with smaller margins.

Since the model begins with a leadership contest within the majority party in Congress and ends with a round of legislative elections, the book will proceed in the opposite order so as to generate equilibrium conditions. The first several chapters will present components of the model, postponing the process of describing equilibria until chapter 4. Chapter 2 will propose a model of voting behavior when voters either re-elect or defeat incumbents on the basis of the roll call votes they cast. Voters will sometimes re-elect incumbents who should be too ideologically extreme to win in standard spatial theory, but they will also reward ideological movement toward the center with at least somewhat higher vote shares. The model will be unidimensional and make strong assumptions about the knowledge and sophistication of the electorate, so the model is supplemented by two appendices: Appendix A will address the issue of dimensionality, and appendix B will provide an alternative model with less sophisticated voters, showing that the predictions hold.

Chapter 3 will move from the final stage of the model (the election) to the penultimate stage (the roll call voting process). As we shall see, even though the electoral punishments for extremism are modest, the policy consequences of insincere roll call voting are zero unless the rest of the chamber is exactly tied on a vote. This emphasizes how puzzling it is that members of Congress incrementally adopt electorally suboptimal positions in the policy space. Even if the electoral costs of extreme positioning are modest, why pay them at all when insincere roll call voting will generally be costless? The rare case of the pivotal legislator will be examined, because it provides insight by pointing out the potentially unboundedly large benefits of policy victories.

Chapter 4 will resolve the remaining puzzles by combining the observations from chapters 2 and 3 into a unified spatial model of legislative elections, parties, and roll call voting, and describe the equilibrium conditions. In the unified model, a majority legislative caucus will first vote for a party/chamber leader, whose platform consists of a disciplinary system. The chamber will then cast roll call votes, constrained by the disciplinary system, and, in the process, legislators incrementally adopt their positions in the policy space. Finally, they face the electoral consequences of the positions that they incrementally adopted. Chapter 4 will derive the conditions under which a legislative caucus demands that leadership solve the collective action problem of sincere voting through the application of preference-preserving influence, and then trace out the implications of that choice for the incremental positioning process and subsequent elections using the results derived in chapters 2 and 3. As we shall see, when the caucus is ideologically homogeneous, electorally diverse, and policy motivated, it will vote for a leader who promises to apply preference-preserving influence. Consequently, caucus members will incrementally adopt ideologically extreme, sincere, and electorally suboptimal positions in the policy space. Doing so will have negative electoral consequences, but those consequences will be mitigated by the mathematical paradoxes derived in chapter 2, thereby explaining why members of Congress incrementally adopt extreme positions and how they win re-election anyway.

Empirical analysis will begin in chapter 5, and the analysis will be divided between pre-and post-2010, since the rise of the Tea Party / Freedom Caucus–dominated Republican majority changed several features of legislative politics, but all of which are compatible with the underlying logic of the model.

Chapter 5 will examine the quantitative historical patterns by which polarization developed in Congress up to the 2010 election. Contrary to conventional wisdom, "centrist" (meaning, competitive) districts have not disappeared over time. They are roughly as common in the early 21st century as they were in the mid-20th. Rather, their representatives have gotten more extreme, tracking their parties' medians rather than their districts as the conditions for party government have developed, putting pressure on leaders to solve the collective action problem of sincere voting.

Chapter 6 will present a case study in solving the collective action problem as Democratic leaders struggled to keep their caucus from voting against the

Patient Protection and Affordable Care Act in 2010, despite electoral incentives to vote no.

Chapter 7 will examine how Congress changed during the divided government period from 2011 through 2016. While the Republican majority did not seem to need party pressure for its members to build ideologically extreme voting records, party leadership evolved to address a different collective action problem. The combination of extreme polarization and divided government introduced by the 2010 election reduced the legislative agenda to little more than bills that avoided ideologically extreme reversion points, exposing previously hidden divisions within the party and altering the direction, but not the fundamental nature, of party pressure.

Finally, chapter 8 will conclude by discussing the future of spatial theory and the inseparability of elections, parties, and roll call voting, as well as the representational consequences of party government in Congress.

2

Voter Preferences over Bundles of Roll Call Votes

The first task in constructing a unified model linking the electoral process to the roll call voting process in Congress is to determine how rational voters would respond to roll call voting records. Then we can derive predictions about how legislators will behave during the legislative session preceding an election based on rational expectations about the electoral consequences of their behavior. As we construct a model of electoral responses to legislative voting records, we will resolve a central puzzle about polarization in Congress: how do ideologically extreme incumbents win re-election given their behavior in office?

As we shall see, depending on the structure of the legislative agenda, rational voters will sometimes prefer an ideologically distant candidate to a more proximate candidate because the bundle of roll call votes implied by a proximate location will not always provide a voter with more utility than the bundle implied by a more distant location, even when the voter has single-peaked, symmetric preferences over *policy*. Most importantly, this "proximity paradox" will occur *because of* the spatial structure of the model, not despite it. In aggregate, then, an electorate will sometimes choose a candidate further from the median voter over a candidate closer to the median voter, even though all voters have single-peaked, symmetric preferences, perfect information, and vote sincerely on the exclusive basis of policy positions. Centrism will provide some electoral advantage, but not a determinative one. Thus, voters may regularly re-elect incumbents who should be too ideologically extreme to win in standard models because conventional approaches to spatial modeling overstate the electoral rewards of centrism.

This is hardly the first effort to explain extremists' electoral success, and there are two simple ways to reconcile that pattern with conventional models. First, we can observe that voters' choices are limited by the candidates. Consider the NOMINATE scale, ranging from −1 to +1. Suppose that the median voter's location is zero. How can an incumbent located at .8 possibly win? The easiest way is if the challenger is even further from the median, say at −.9. Given those candidate locations, a Downsian model of voting behavior would predict victory by the candidate located at .8, even though that candidate is an extremist. Empirical analysis of challengers' positions, such as Ansolabehere et al. (2001), has demonstrated that challengers are rarely more centrist than their partisan compatriot officeholders, which helps explain why unrepresentative extremists of one party are frequently replaced by unrepresentative extremists of the other (Bafumi & Herron 2010). In principle, then, little may be amiss with voters. However, there are two problems with this defense of conventional spatial theory. First, it simply shifts the problem from voters to candidates. If an incumbent located at .8 defeats a challenger located at −.9, that makes sense from a Downsian perspective on *voters*, but not from a Downsian perspective on *candidates* because the question becomes why the challenger located at −.9 didn't just run on a platform at least some small distance to the right of −.8, thereby guaranteeing victory? After all, it would require relatively little movement. Pointing to a more extreme *ex*ogenous location does not solve the problem if those exogenous locations are at odds with what would seem to be equilibrium predictions when the locations are *end*ogenous.

The second problem is that while empirical analysis does show that candidate positioning affects incumbent vote shares in the *direction* that spatial theorists predict (see, for example, Ansolabehere et al. 2001; Ansolabehere & Jones 2010; Bovitz & Carson 2006; Brady et al. 2007; Canes-Wrone et al. 2002; Carson et al. 2010; Erikson & Wright 2000), which would be compatible with the idea that voters are simply constrained by candidate choice, the magnitude of the effect is never overwhelming. Centrism conveys at least a modest electoral advantage, but hardly a determinative one. Voters select an extremist over a comparatively centrist alternative with surprising frequency, so even though there is a general tendency for centrists to do better than extremists, deviations from that tendency are too great for complacency in spatial theory.

The second simple approach to defending spatial models of voting behavior is to ask, what if we are simply drawing district lines in a way that leaves district median voters relatively far from the absolute center? If Democrats represent districts with left-leaning medians and Republicans represent districts with right-leaning medians, then districts are electing extremists who reflect their own collective preferences. While this explanation is popular among journalists and pundits, scholarship has been less than kind to the notion. Redistricting plays at most a very minor role in legislative polarization (Abramowitz et al. 2006; Brunell 2006, 2008; Buchler 2005, 2009, 2010a, 2010b; McCarty et al. 2009), and chapter 5 will demonstrate that representatives from competitive districts track their party medians rather than their district medians over time. This creates a significant problem for spatial theory because it exhausts the last of the simple methods to reconcile voters' willingness to vote for extremists with "pure" Downsian theory.

Consequently, most attempts to reconcile legislative extremism with spatial theory have added something to the Downsian, unidimensional spatial model so as to dilute the effect of extremism along the liberal-conservative spectrum. There is one way to do so that does not abstract away from spatial theory itself—adding other policy dimensions. Despite public and scholarly attention to the liberal-conservative divide, neither the electorate nor even Congress at its most polarized can be described fully with a unidimensional spatial model. While the first dimension of Poole and Rosenthal's NOMINATE scores can explain more than 90% of all roll call votes in the modern era, that isn't 100%. The problem comes from McKelvey's (1976) observation, aided by Schofield (1978), that without pure unidimensionality, many predictions of spatial models break down. In a multidimensional space, there is no unique median voter, in which case the median voter theorem does not apply.

As an explanation for the electoral success of left-right extremists in the modern Congress, though, additional policy dimensions are less than compelling. In a multidimensional policy space, there is no Condorcet-winning location, so, by definition, no incumbent's location can be a Condorcet winner. There must be at least one other location in the policy space that would defeat the incumbent's location, which is fixed by past voting records. Thus, a challenger unconstrained by past votes should be able to defeat the incumbent. Rather than overwhelming incumbent re-election rates, we would observe overwhelming incumbent loss rates as challengers exploit intransitivities in collective preferences. Challengers, not incumbents, would be at an advantage. If one's objective is to explain changes over time in the ideological composition of the Democratic Party as it transitions away from a southern base, then unidimensional models are insufficient. However, additional policy dimensions provide a poor explanation for the electoral success of left-right extremists in the modern context. This chapter will focus on the unidimensional case, then, for reasons of both pragmatism and parsimony. Nevertheless, the central insights in this chapter will be applicable to multidimensional models as well, and appendix A will elaborate on such models.

Turning away from the policy space itself, other models have included nonpolicy factors to explain the electoral success of extremists. One of the most prominent methods has been to build a model in which voters are motivated by some combination of policy and nonpolicy considerations. Stokes (1963) suggested that voters have preferences over a combination of positional issues and valence issues. Positional issues are those, such as abortion, over which voters have different preferences over policy outcomes themselves. Valence issues, though, are those over which voters agree about objectives but disagree about who is best able to achieve those objectives. Spatial theorists have adapted the concept of the valence "issue" and transformed it into a valence "characteristic," which is a candidate trait that voters intrinsically value, such as competence or honesty. Including voter preferences over these characteristics can allow rational voters to prefer distant candidates over proximate candidates as long as the distant candidate has a valence advantage, and that process can alter the predictions of spatial models (see, for example, Enelow & Hinich 1982; Cukierman 1991; Londregan & Romer 1993; Adams 1999; Ansolabehere & Snyder 2000; Roemer 2001; Groseclose 2001; Schofield 2004; Adams et al. 2005; Kartik & McAfee 2007; Schofield & Sened 2005a, 2005b). Such factors can potentially explain incumbents' overwhelming re-election rates, despite their extremism.

The problem is that if we include such nonpolicy characteristics in our spatial models in order to explain the failure of proximity-based voting, then we are explaining nearly all of legislative election results with the components of the model unrelated to the policy space because it is so rare for a less experienced candidate to defeat a more experienced candidate. At that point, it is not clear that the policy space in the model adds much of anything. It becomes the equivalent of the stones in "stone soup." In the fable of stone soup, a man promises a group of people that he knows the recipe for the best soup in the world—stone soup, in which the key ingredient is a set of stones. He will prepare the stone soup, and supply the critical ingredient (the stones), but the others will need to provide a few other secondary ingredients, such as the meat, vegetables, and herbs. Lo and behold, the soup is delicious, but the con is that the stones themselves have nothing to do with that. The actual nutrition and flavor come from the ingredients that the man convinced the others to provide. If we explain the electoral victory of ideological extremists with voters' preference for nonpolicy characteristics, like experience, then the spatial policy component of the model is no more useful for explaining outcomes than the stones are to stone soup. In contrast, this chapter will show how we can generate deviations from proximity voting from the policy space itself rather than from an orthogonal "valence" dimension. This is not to say that experience is unimportant. Empirically, we must accept that candidate experience plays a critical role in legislative elections because voters demonstrably have strong preferences over experience in legislative elections. The purpose here is to argue that we do not *need* experience to explain widespread violations of proximity voting. The experience effect is not "crowding out" ideology—it is filling in a gap left by a relatively weak ideological effect.

An even more simple approach to explaining violations of proximity voting would be to incorporate imperfect information. If voters do not know the candidates' locations in the policy space with perfect certainty, or if their perceptions are influenced by factors such as partisanship, then a noncentrist can occasionally defeat a centrist because of what is essentially voter error.

Adams et al. (2005) explored the role of partisan bias in spatial theory, and more recently, Jessee (2012) has proposed a psychologically motivated spatial model in which ideological perceptions are shaped by partisanship. Similarly, Sniderman and Stiglitz (2012) argue that party reputations condition the effects of candidates' positions on voting behavior. As with valence, each of these models taps into a real and important phenomenon, but if we explain the failure of the central prediction of spatial theory by including a factor that either works against or distorts the spatial effect, then it is not clear what spatial theory itself contributes to the discussion, particularly when polarization is the historical norm rather than the exception. This is a particular problem for spatial theory if partisanship is the explanation, though, because partisanship plays such an important empirical role in voting behavior that if partisanship-induced errors explain non-Downsian election results, then the errors must be so common that it is unclear what we gain by focusing on what the "correct" decisions would have been. Ultimately, we would still be arguing that extremists win re-election because voters just don't vote spatially, even if they are trying to do so.

An alternative approach to explaining violations of proximity voting has been to abandon the assumption that voters always vote sincerely. In elections with more than two candidates, strategic voting is almost necessary because of Duverger's law, but even in two-candidate elections, strategic voting can be an analytically useful assumption. Such models are the few places in which spatial theorists take seriously the postelection legislative process. Consider, for example, Austen-Smith (1984, 1986), who models postelection coalition formation within a legislative body, or from an empirical perspective, consider Lacy and Paolino (1998), who address the importance of the bargaining process across different branches for strategic voters. Making the Lacy and Paolino argument more complicated, Sin (2014) argues that legislative procedures themselves are influenced by interbranch relations. All of these issues may be of importance to strategic voters.

Strategic voting models do not reduce the role of spatial theory to that of the stones in stone soup, nor do they rely on some source of voter error. Instead, the method by which these models generate violations of proximity voting is fundamentally spatial in nature. These arguments are built around the idea that voters who vote for a distant candidate over a more proximate candidate are, in some sense, "playing a deeper game" than a simple Downsian model. They are looking down the line a few moves and determining that what seems like a victory in the short run may turn into a defeat a few moves later. Electing a more proximate legislative candidate may result in a coalition forming that will produce a more distant policy outcome.

The problem with this general approach to modeling is more subtle. How do you pick a stopping point? We cannot simply observe a strategic choice that seems suboptimal and keep adding stages to the game until we find one at which it appears optimal. Suppose that a strategic decision leads to a suboptimal outcome at time t but an optimal outcome at time $t + 1$. We cannot end the model there and declare the decision optimal. The problem is, what about time $t + 2$? Suppose, for example, that an electoral sacrifice at time t leads to a policy victory at time $t + 1$, but that leads to an electoral defeat at time $t + 2$, which has negative policy consequences. Adding stages to the model until we find one in which the behavior we observe appears optimal comes dangerously close to losing a coin toss and then demanding best two out of three, and upon losing that toss, demanding best three out of five, and so on. Thus, even the strategic voting models leave something to be desired as far as explaining violations of proximity voting in spatial theory.

Moreover, the modeling approaches discussed above are all built on the same substantive mischaracterization of the policymaking process. At the end of either the election or the postelection legislative wrangling, each of these sets of models characterizes the eventual outcome as the selection of a single policy location in the policy space, and voters seek to minimize the distance between this unitary outcome and their own ideal points. However, a central motivating principle of this book is the observation that this ignores the substantive positioning process. We cannot characterize the policy outcome of a legislative election with a unitary point because the primary responsibility of a legislator is to cast a sequence of roll call votes, thereby incrementally adopting a position in the policy space. So the model in this chapter is built around a truly simple question—what if voters have preferences, not over the candidates' policy locations, but over the bundles of roll call votes implied by those locations and, indeed, used to *construct* those locations? The central result will be the potential for what this book will call a "proximity paradox," in which voters with single-peaked, symmetric preferences and perfect information sincerely prefer the bundle of roll call votes associated with a distant candidate location to the bundle associated with a more proximate location. Thus, violations of proximity voting follow naturally from the policy space without adding nonpolicy parameters to the model, without allowing mistakes, without adding new stages to the model until actions that initially seem irrational eventually appear rational, and without abandoning the Downsian framework. After all, the roll call voting process is not one that *follows* from the location process. It *is* the positioning process. If violations of proximity voting follow from that process, then we can explain the electoral success of extremists in the context of spatial theory, even unidimensional spatial theory, simply by linking spatial models of elections to spatial models of roll call voting. These violations of proximity voting are a *quirk* of unidimensional spatial politics, not a *deviation* from spatial politics.

An Alternative Spatial Model of Voting Behavior

This chapter will present a unidimensional spatial model of voting behavior that is conventional in every way except two: the mechanism by which legislators provide utility to voters, and the sequence. Let us begin with the conventional. Voters will have quadratic loss utility functions for policy within a unidimensional policy space. They will select a candidate in a two-candidate election with perfect information.

The use of quadratic loss utility functions reflects the common assumption of risk aversion, rather than the risk neutrality assumed by absolute loss utility functions. Proximity paradoxes will result from two factors to be discussed in this chapter—asymmetric frequencies and asymmetric differentials. While the latter will depend partially on risk aversion, the former will not, so the proximity paradox will not be dependent on risk aversion.

Next, consider dimensionality. The model here will focus on the unidimensional case for both conceptual and practical reasons. While it is sometimes possible to generate equilibria in a multidimensional spatial model, it usually requires a "trick," such as the inclusion of valence characteristics, as in Ansolabehere and Snyder (2000). The goal here is to generate violations of proximity voting without including such factors. Moreover, if additional dimensions are not the cause of extremists' electoral success, then they add complication without obvious benefit. Nevertheless, the principles of the proximity paradox will work equally well in a multidimensional context, as discussed in appendix A.

The use of perfect information in a two-candidate race similarly avoids the possibility of voter error or the necessity of strategic voting, which can generate violations of proximity voting. In fact, the model will assume that voters are hypercompetent, in terms of both how informed they are and how they approach calculations. The point is not that voters are this sophisticated, but that we need not resort to voter error or lack of sophistication to explain nonproximity voting. Nevertheless, appendix B will present a related model with similar predictions when voters are less sophisticated.

This chapter will, however, treat the candidates' locations as exogenous. The purpose of this chapter is to derive predictions about voting behavior and use those predictions in subsequent chapters to generate predictions about how incumbents behave in office, in anticipation of the electoral consequences. Equilibrium conditions, then, will be derived more fully in chapter 4.

The model will break from the conventions of spatial theory in two respects: the sequence, and the mechanism by which voters receive utility. Conventional spatial models assume that the winner implements her campaign platform after the election, but, as discussed in the introduction, this model reverses the order. Consequently, voters will retrospectively evaluate the incumbent in comparison to what the challenger would have done had she been in office at the time. Toward the end, discussion will address the implications of the mathematical principles for a prospective voting model.

The most important break from convention, though, will be in the mechanism by which voters receive utility. In a traditional spatial model, if a candidate wins a legislative election on a policy platform of p^*, that candidate simply provides all voters in her district with a policy equivalent to p^* in a single action after the election. However, that single action does not resemble any real action ever taken by a legislator. A sitting legislator casts a series of roll call votes. Each roll call vote, j, is a choice between a specific status quo point, which can be represented by sq_j, and a specific alternative point, which can be represented by a_j. For a legislator, having a location of p^*, or rather, incrementally adopting a location of p^*, means that for any choice between sq_j and a_j, the legislator will vote for sq_j if $|p^* - sq_j| < |p^* - a_j|$, and the legislator will vote for a_j if $|p^* - a_j| < |p^* - sq_j|$. So a sitting legislator whose location can be characterized by p^* does not provide her constituents with a policy of p^*. Instead, a legislator *adopts* a location of p^* by providing her constituents with a bundle of roll call votes implied by a location of p^*. So the model of voting behavior will be one in which voters have quadratic loss utility functions for policy, and rank candidates not by their preferences over candidates' locations, but by their preferences over the bundles of roll call votes through which legislators adopt those locations.

A point that will become noteworthy later, though, is that this chapter will assume voters derive utility from both pivotal and nonpivotal roll call votes equally. There are both practical and theoretical reasons. On the practical level, very few roll call votes are decided by a single vote in Congress, so if voters only responded to pivotal roll call votes, they would frequently have nothing to evaluate, and we would have nothing to model. On a theoretical level, though, the logic behind retrospective evaluation is that it is a tool for prospective assessment. A bundle of roll call votes, even nonpivotal ones, is an indication about what a legislator would do in the future if she turned out to be pivotal. Thus, nonpivotal roll call votes deserve consideration too. However, these issues will be revisited in chapters 3 and 4 because there are compelling reasons to have members of Congress make a distinction between pivotal and nonpivotal roll call votes.

Voters' Utility Functions

Suppose that a legislative session consists of a set of roll call votes where each roll call vote, j, consists of a specific status quo point, sq_j, and a specific alternative point, a_j. This set can either represent the complete set of roll call votes, or merely a subset of important and publicly salient roll call votes, factoring out the low-profile, nonideological votes. This subset may be the subset given attention by the media and campaigns, but it is important to keep in mind that the model will work equally well regardless of whether or not voters consider the complete legislative agenda. Incumbent legislators will draw attention to the agenda items on which their votes were consistent with constituent preferences, and challengers will draw attention to those on which the incumbent's vote was inconsistent with constituent preferences. For example, Democratic incumbents facing re-election in 2010 called attention to votes for infrastructure spending within their districts to aid recovery from the "great recession," while their Republican challengers called attention to the incumbents' votes on healthcare reform. Each bill also implies a status quo—the absence of stimulus spending in the former case, and the prior healthcare system in the latter. Campaign dynamics, then, create subsets of roll call votes that policy-motivated voters can evaluate, and the arithmetic will be identical regardless of whether voters examine the complete legislative agenda or merely a subset highlighted by the campaign.

A policy platform given by a location of p^* does not provide voters with a policy equivalent to p^*. Instead, it provides voters with a bundle of roll call votes implied by that location given the sequence of roll call votes. So, let $\{p^*_1, p^*_2, \ldots, p^*_k\}$ represent the sequence of k policy options for which a legislator or potential legislator would vote in a sequence of k roll call votes given a platform represented by a location of p^*. In this array, $p^*_j = sq_j$ if $|p^* - sq_j| < |p^* - a_j|$, and $p^*_j = a_j$ if $|p^* - a_j| < |p^* - sq_j|$.

Suppose that an arbitrary voter has a utility function for policy, p, given by $U(p) = -(i - p)^2$ where i is the voter's ideal point. This utility function is both single-peaked and symmetric. However, a candidate with a platform of p^* does not provide voters with a policy of p^*. So the voter will not ascribe a utility value of $-(i - p^*)^2$ to that platform. Instead, the utility that the voter will ascribe to that platform is the sum total of all utility she ascribes to the roll call votes implied by p^* given the sequence of k roll call votes. Thus, the voter's utility for electing a candidate with a platform of p^* is given by the following series: $\sum_{j=1}^{k} -\left(i - p^*_j\right)^2$, where p^*_j is defined as above.

Hence, when faced with a choice between an incumbent whose location is A and a challenger whose location is B, the voter will not simply choose whichever candidate is located closest to her ideal point in the policy space. Instead, let $\{A_1, A_2, \ldots, A_k\}$ represent the sequence of roll call votes that the incumbent did cast, and let $\{B_1, B_2, \ldots, B_k\}$ represent the sequence of roll call votes that the challenger *would have cast had she faced the same legislative agenda*. A voter who considers the incremental process by which legislators actually position themselves will vote for the incumbent, A, if $\sum_{j=1}^{k} -\left(i - A_j\right)^2 > \sum_{j=1}^{k} -\left(i - B_j\right)^2$, and she will vote for the challenger, B, if $\sum_{j=1}^{k} -\left(i - A_j\right)^2 < \sum_{j=1}^{k} -\left(i - B_j\right)^2$.

Preference Orders and Bundles of Roll Call Votes

Constructing an agenda in which voters sincerely prefer a bundle of roll call votes associated with a distant legislator location to the bundle associated with a more proximate location given the same agenda is remarkably easy, and agendas that make substantive sense given actual legislative politics can create this effect. Consider a voter with an ideal point, *i*, facing a choice between an incumbent member of Congress and a challenger whose locations are given by MC and C respectively. As mentioned earlier, we will treat these locations as fixed and exogenous here, and in chapter 3, we will begin to examine the incumbent's roll call voting process as a strategic process based on the predictions here, and then develop equilibrium conditions in chapter 4.

Suppose that $MC < i < C$, and that $i - MC > C - i$. So the voter's ideal point is located between the two candidates' positions, and the challenger's platform is closer to the voter's ideal point than the member of Congress's position. If the voter calculates the utility she derives from a legislator's term in office in the traditional manner, she will prefer the more proximate candidate—the challenger. However, if she derives utility from the roll call votes that must be cast in order for either candidate to actually adopt his or her respective locations in the policy space, she may sincerely prefer the more distant candidate—the incumbent member of Congress. When that happens, we have what may be called a "proximity paradox" (Buchler 2011a), and such a paradox can permit ideologically extreme incumbents to defeat more moderate challengers, thereby resolving the first major puzzle of the book.

Legislative Agendas Producing a Proximity Paradox

Whether or not voters practice proximity voting given the utility function specified above depends on the structure of the legislative agenda. Essentially, certain types of legislative agendas can induce voters whose utility functions for *policy* locations are symmetric to behave as though their preferences for *candidate* locations are *a*symmetric. We can distinguish between two forms of asymmetry in the legislative agenda capable of producing a "proximity paradox": asymmetric frequencies and asymmetric differentials. An agenda with asymmetric frequency is an agenda in which the number of roll call votes with cut points in a given region of the policy space varies across the policy space. Alternatively, an agenda with asymmetric differentials is an agenda in which the roll call votes with cut points on one side of a voter's ideal point are associated with greater utility differentials between the status quo and alternative points than the roll call votes with cut points on the other side of a voter's ideal point. Consider the elaborations below.

Each roll call vote, *j*, is a choice between a status quo point, sq_j, and an alternative point, a_j. Without loss of generality, let us assume that $sq_j > a_j$. For any such choice, there exists a cut point given by $(sq_j + a_j) / 2$, such that any actor with an ideal point to the left of $(sq_j + a_j) / 2$ will prefer the alternative point, and any actor with an ideal point to the right of $(sq_j + a_j) / 2$ will prefer the status quo point. An agenda with "symmetric frequency" is one in which the number of roll call votes with cut points in a given region is a function of the size of that region. For example, if any given region of the policy space of size *d* has approximately *n* roll call votes with cut points in that region, then we have essentially a linear relationship. A region of size 2*d* will have 2*n* roll call votes with cut points in that region, regardless of where that region is. Symmetric frequency will occur if, for example, cut points are randomly distributed across the policy space. However, an agenda with "asymmetric frequency" occurs when the number of roll call votes with cut points in a region of a fixed size, *d*, varies across the policy space. Consider the implications of an agenda with asymmetric frequency on voters' decisions.

Suppose that a voter's ideal point is 5, the incumbent member of Congress is located at 2, and the challenger's position is 7. In the most simple example of asymmetric frequency, suppose that the legislative agenda includes a single vote with a cut point in the interval of (5, 7), and no roll call votes with cut points in the interval of (2, 5). Thus, despite the fact that the challenger's location is closer to the voter's ideal point than the incumbent's location, the incumbent will have cast no roll call votes at odds with the voters' preferences, but the challenger would have cast a roll call vote with which the voter disagrees—the vote with a cut point in the (5, 7) interval. Thus, the voter must prefer the bundle of roll call votes implied by the incumbent's location and, in fact, cast in order to adopt that location, over the bundle of roll call votes implied by the challenger's location, even though the challenger's location is closer to the voter's ideal point. Hence, we have a simple proximity paradox generated by asymmetric frequency.

However, that example is based on a single roll call vote. One might argue that the distance between the incumbent's position and the voter's ideal point is irrelevant if there are no roll call votes with cut points between their locations. In fact, we might argue that this type of agenda makes the voter's ideal point and the incumbent's position functionally indistinguishable, in which case the proximity paradox is uninteresting, if not a downright cheat. Moreover empirically, the incumbent's position and the voter's ideal point would appear to be the same since we primarily estimate legislators' locations by the votes that they cast in order to adopt their locations anyway. Thus, this simple example of a proximity paradox is uninteresting for the purposes of the argument here. However, there is no necessary upper bound on the total number of roll call votes in any given range. In principle, there could be 200 roll call votes with cut points in the (5, 7) range and only five in the (2, 5) range.

So suppose that the difference between the utility a given voter receives from the status quo and alternative points is fixed at some value, z, for all roll call votes, meaning that the agenda is characterized by symmetric differentials: $|U_V(sq_j) - U_V(a_j)| = z$ for all j. In that case, the voter's preference between a candidate located at 2 and a candidate located at 7 depends entirely on which interval contains more roll call cut points: (2, 5) or (5, 7). If there are m roll call votes with cut points in the (2, 5) interval, and n roll call votes with cut points in the (5, 7) interval where each roll call is associated with a fixed utility differential of $z > 0$, then the cost of the roll call disagreements implied by a candidate location of 2 is mz, and the cost of the roll call disagreements implied by a candidate location of 7 is nz. Thus, if $m < n$, the total utility of the bundle of roll call votes associated with a location of 2 must be greater than the total utility of the bundle of votes associated with a location of 7, even though 7 is closer to the voter's ideal point. Hence, we have a more interesting proximity paradox generated by asymmetric frequency. A voter with single-peaked, symmetric preferences over *policy* prefers a distant *candidate* to a more proximate *candidate*, even when the structure of the agenda permits us to observe the distances, because legislators do not provide their constituents with their policy platforms in a single action—they provide bundles of roll call votes associated with their locations in the policy space.

However, there is another method of generating a proximity paradox that will turn out to be more analytically useful here—the potential for an asymmetric differential. An asymmetric differential occurs when a voter ascribes different "stakes" to different roll call votes. Continuing with a voter whose ideal point is 5, an incumbent located at 2 and a challenger located at 7, consider the following hypothetical legislative agenda. For the purposes of this example, we will set the initial status quo point for all policy issues at 6.5 to show that the process *can* work when initial status quo points are constrained to the same location. In a later demonstration, we will consider the potential for a nonuniform distribution of initial status quo points. So consider an agenda consisting of six bills across six different policy areas, each moving policy within one specific issue from an initial status quo of 6.5 to some other alternative location (Table 2.1).

The agenda is constructed in such a way that policy moves toward the incumbent member of Congress because we are developing toward a model focused

Table 2.1 **An agenda with asymmetric differentials and uniform initial status quo points**

Bill	*Issue*	*Status quo*	*Alternative*	*Cut Point*
1	Taxes	6.5	3	4.75
2	Abortion	6.5	5	5.75
3	Immigration	6.5	5.5	6
4	Environment	6.5	3.25	4.875
5	Defense $	6.5	3.25	4.875
6	Welfare $	6.5	3.25	4.875

on a majority party caucus in the House of Representatives, and this should reflect policy movement toward majority party preferences. In this six-vote legislative agenda, there are four votes on which the member of Congress, located at 2, casts a vote that conflicts with the voter's preferences. In contrast, there are only two votes on which the challenger, located at 7, would have done so. In this case, then, the asymmetric frequency of roll call votes actually favors the challenger, thereby stacking the deck *in favor* of proximity voting. However, the asymmetric differentials favor the incumbent. So consider how a voter with a utility function given by $U(p) = -(5 - p)^2$ will rank a legislator position of 2 and a legislator position of 7, given this agenda. A legislator located at 2 would vote for the alternative point on all six bills because the status quo location of 6.5 is further from 2 than the alternative locations of 3, 5, 5.5, 3.25, 3.25, and 3.25. The voter's utility for those votes is as follows.

$$U(3)+U(5)+U(5.5)+U(3.25)+U(3.25)+U(3.25)=-(5-3)^2-(5-5)^2 -(5-5.5)^2-(5-3.25)^2-(5-3.25)^2-(5-3.25)^2=-13.4375 \quad (2.1)$$

On the other hand, consider the voter's utility for the votes that would be cast by a legislator located at 7. A legislator located at 7 would vote for the status quo option on each bill because the distance between 7 and each status quo point is fixed at 0.5, but the distances between each status quo point and the alternative points are 4, 2, 1.5, 3.75, 3.75, and 3.75 respectively. The voter's total utility from having a legislator vote against each bill is as follows.

$$U(6.5)+U(6.5)+U(6.5)+U(6.5)+U(6.5)+U(6.5)=-(5-6.5)^2-(5-6.5)^2 -(5-6.5)^2-(5-6.5)^2-(5-6.5)^2-(5-6.5)^2=-13.5 \quad (2.2)$$

Given this hypothetical legislative agenda, the voter would derive a utility of −13.4375 from the bundle of votes cast by the incumbent, and a utility of −13.5 from the bundle of votes that would have been cast by the challenger. Thus, the voter will still prefer the bundle of votes offered by incumbent member of Congress to the bundle of votes offered by the challenger. Consider the oddity of this preference ordering. The voter's ideal point is further from the incumbent's position than it is from the challenger's position, so by proximity, we would expect the voter to prefer the challenger. Furthermore, there are more roll call votes on which the incumbent votes against the voter's preferences than roll call votes on which the challenger would have done so. Thus, the asymmetric frequency of the agenda also favors the challenger. However, the votes that the challenger *would have* cast against the voter's preferences are more important to the voter than the votes that the incumbent *did* cast against the voter's preferences. The utility differential between the challenger's votes and the incumbent's votes on Bills 2 and 3 are 2.25 and 2, respectively, whereas the utility differentials between the incumbent's and challenger's votes on Bills 4 through 6 are only 0.8125 each. Thus, the bundle of votes implied by the incumbent's location, and cast in order to adopt that location, is more appealing to the voter than the bundle of votes that the challenger would have needed to cast in order to adopt her location. When voters evaluate legislators by the roll call votes they cast in order to adopt their positions in the policy space, they will sometimes prefer a distant *candidate* to a proximate *candidate,* even when they have single-peaked, symmetric preferences over *policy*. When that occurs, we have a "proximity paradox."

The Unidirectional Proximity Rule

There is a limitation on the possibility of a proximity paradox, and it will have an important conditioning effect on the ability of the proximity paradox to explain polarization. In order for the proximity paradox to occur for any given voter, the two candidates must be on alternative sides of the voter's ideal point. This is "the unidirectional proximity rule." Consider any arrangement of a voter's ideal point, i, and two arbitrary candidate locations, A and B, such that $i \leq A < B$. The voter's preference ordering between these two candidates depends on whether or not there is at least one roll call vote, j, such that $A < (sq_j + a_j) / 2 < B$. If there are no such roll call votes, then the voter is indifferent between A and B because the positions are substantively identical given the legislative agenda, that is, they would cast the same bundle of votes in order to adopt their locations given the agenda. However, if there is at least one roll call vote such that $A < (sq_j + a_j) / 2 < B$, then on this roll call vote, A will cast a different vote from B, and the voter will prefer A's vote to B's vote because the voter and A are on the same side of the cut point and B is on the opposite side. Moreover, there can be no roll call votes such that A and B would cast different votes, but the voter prefers B's vote to A's vote. This is because in order for A and B to cast different votes on a bill, the cut point for the bill must be within the interval (A, B), but if that is the case, then the voter agrees with A's vote, but not B's vote. Thus, if A's position is substantively distinguishable from B's position given the legislative agenda, a voter with a location of i must prefer A to B if $i \leq A < B$. Similarly, a voter with a location of i where $A < B \leq i$ must prefer B to A if A and B are distinguishable given the agenda. A proximity paradox can only occur for a given voter when the two candidates are on opposite sides of that voter's ideal point. As we shall see later, this limitation will create an important role for parties.

Individual Preferences and Aggregate Election Results

While the unidirectional proximity rule puts an important limitation on violations of proximity voting, the critical swing voters in any given election are likely to have ideal points located between the two competitors' locations, so the potential for a proximity paradox still has important implications for aggregate election results. If an individual voter with single-peaked, symmetric preferences over policy can prefer a distant candidate to a proximate candidate, then it is trivially easy to show that an electorate consisting of voters with single-peaked, symmetric preferences over policy can prefer a candidate located further from the median voter's ideal point to a candidate located closer to the median voter's ideal point.

Consider an extreme but particularly useful hypothetical. Suppose that an electorate consists entirely of voters with identical utility functions. If a proximity paradox exists for *one* such voter, the same paradox exists for *all* such voters. Thus, the entire electorate will vote for the distant candidate over the proximate candidate in the presence of a proximity paradox. This hypothetical is particularly informative because it reflects the behavior of a homogeneous electorate, even a homogeneously centrist electorate.

This demonstration sheds valuable light on our first central puzzle—can we reconcile the electoral success of ideological extremists with spatial theory without resorting to tactics that dilute the spatial components of our model? Yes, for a very surprising reason. Just because voters prefer the most proximate *policy* doesn't mean that they prefer the most proximate *legislator*, because legislators adopt their positions incrementally through the roll call voting process, so voters' preferences over *legislators'* locations depends on the legislative agenda, which can essentially "warp" the policy space.

However, there is a caveat, which will set the stage for a critical puzzle in chapter 3. Just because the candidate closest to the median voter's ideal point might lose, that doesn't mean there are no incentives to move toward the median voter's ideal point. Suppose that an electorate consists of voters with single-peaked, symmetric preferences over policy, with a unique median voter whose ideal point is i_{MV}. Suppose, further, that the incumbent's position in the policy space is MC and that the challenger's position is C, where MC < i_{MV} < C, and i_{MV} – MC > C – i_{MV}. Thus, the incumbent member of Congress and the challenger are on alternative sides of the median voter's ideal point, with the challenger's position closer to the median voter's ideal point than the incumbent's position. As we have already seen, whether the electorate prefers the incumbent or challenger depends on the legislative agenda, so the challenger's relative centrism does not guarantee victory. However, what would happen if the incumbent member of Congress were to adopt a more "centrist" position of MC* where MC < MC* < i_{MV}?

Adopting a position of MC* rather than MC requires casting a slightly different bundle of roll call votes. For any roll call vote, *j*, such that MC < $(sq_j + a_j) / 2$ < MC*, adopting a position of MC* requires casting a different vote than adopting a position of MC. What are the consequences of any such change for the preferences of the electorate? Let us segment the electorate into four intervals, **1:** (−∞, MC), **2:** (MC, MC*), **3:** (MC*, C) and **4:** (C, ∞).

Interval 1: (−∞, MC). Any voter with an ideal point in the range of (−∞, MC) will prefer the bundle of votes needed for the member of Congress to adopt a position of MC to the bundle of votes needed for the challenger to adopt a position of C, but any such voter will also prefer the bundle of votes needed to adopt a position of MC* to the bundle of votes needed to adopt a position of C. The reason is simple. If *i* is in Interval 1, it follows that *i* < MC < MC* < C. By the unidirectional proximity rule, the voter must prefer MC to C, and she must prefer MC* to C. Thus, the incumbent's decision to adopt a more "moderate" position of MC* rather than MC cannot affect the preference orderings of voters with ideal points in Interval 1. They will still prefer the incumbent, somewhat more grudgingly.

Interval 2: (MC, MC*). For voters whose ideal points are located between the incumbent's modified and unmodified positions, we must break the process down to two questions: how would the voter rank the incumbent's unmodified position relative to the challenger's position, and how would the voter rank the incumbent's modified position relative to the challenger's position? We will begin with the first question. What can we say about the preferences of a voter in Interval 2 over the challenger's position and the incumbent's unmodified position? Without the structure of the legislative agenda, nothing. Such a voter might prefer the bundle of votes associated with MC to the bundle of votes associated with C, but she also could prefer the bundle of votes associated with C to the bundle of votes associated with MC, regardless of which position is closer to her ideal point. That is the indeterminacy result derived earlier by the possibility of a proximity paradox.

However, we *can* derive a prediction about such a voter's preference ordering between MC* and C. Since any voter with an ideal point located in Interval 2 will have an ideal point, *i*, such that $i < MC^* < C$, by the unidirectional proximity rule, any such voter *must* prefer the bundle of roll call votes implied by a location of MC* to the bundle of roll call votes implied by a location of C if MC* and C are substantively distinguishable. Thus, the decision to adopt a position of MC* rather than MC cannot *decrease* the number of votes the incumbent receives from voters in Interval 2, but it *can* increase the number of votes the incumbent receives from such voters.

Interval 3: (MC*, C). Because of the potential for a proximity paradox, voters with ideal points in Interval 3 could potentially prefer MC to C, or they could prefer C to MC. Similarly, they could prefer MC* to C, or they could prefer C to MC*. Despite that, we can still say something about how such voters would respond to the incumbent's decision to adopt a location of MC* rather than a position of MC. For any roll call vote, *j*, such that $MC < (sq_j + a_j) / 2 < MC^*$, adopting a position of MC* requires voting for the more conservative option, whereas adopting a position of MC requires voting for the more liberal option. Moreover, roll call votes with cut points in this range are the only votes on which adopting a location of MC* requires casting a different vote from adopting a position of MC. Since a voter with an ideal point in Interval 3 prefers the more conservative option on any such vote, adopting a location of MC* yields all voters with ideal points in Interval 3 more utility than adopting a location of MC. Thus, it is not possible for the incumbent member of Congress to *decrease* her vote share among voters in Interval 3 by adopting a position of MC* rather than MC, but it is possible for her to *increase* her vote share among these voters by doing so.

Interval 4: (C, ∞). Any voter with an ideal point in Interval 4 will prefer the challenger to the incumbent member of Congress regardless of whether the member of Congress modifies her position or not. If *i* falls within Interval 4, it follows that $MC < MC^* < C < i$, so by the unidirectional proximity rule, if C and MC* are empirically distinguishable, the voter must prefer C to MC, and she must prefer C to MC*. Thus, voters in Interval 4 cannot have their preference orderings affected by the incumbent's decision to adopt a position of MC* rather than MC.

Combining our predictions about voters in Intervals 1 through 4 allows us to make a general point about the electoral effects of the incumbent deciding to adopt a more moderate position of MC* rather than MC—it cannot cause any voter who would have supported her with a position of MC to abandon her, but it can attract supporters who otherwise would have preferred the bundle of roll call votes implied by a position of C. Adopting a position of MC* rather than MC, then, will tend to increase the incumbent's vote share.

In fact, though, it is not simply movement toward the *challenger* that benefits the incumbent. The position of the median voter's ideal point is a Condorcet winner, just as in more conventional models. Suppose that the location of the unique median voter is i_{MV}. Suppose that there are two candidates located respectively at A and B, where $A = i_{MV}$ and $B > i_{MV}$. Suppose, further, that there is at least one roll call vote, *j*, such that $A < (sq_j + a_j) / 2 < B$, thus ensuring that A and B are functionally distinguishable positions. By the unidirectional proximity rule, any voter with an ideal point to the left of the median voter must prefer the bundle of votes implied by a position of A to the bundle of votes implied by a position of B. Moreover, since A's position is identical to the median voter's ideal point, a position of A implies *always* voting in accordance with the median voter's preferences, but a position of B does not, so the median voter must also prefer the bundle of votes used to adopt a position of A to the bundle of votes used to adopt a position of B. Thus, more than 50% of the electorate must prefer a legislator with a position of A to a legislator with a position of B. Thus, the location of the median voter's ideal point cannot be defeated by any alternative position in the policy space. So, while we have seen that the incremental positioning process can create strange preference orderings, two critical results from conventional spatial theory remain applicable—the median voter's location is the electorally optimal position, and voters will vote for the most proximate candidate whenever the two candidates' locations are on the *same side* of the voter's ideal point. However, if the candidates deviate even slightly from a voter's ideal point, in alternative directions, voters can prefer distant candidates to proximate candidates, allowing extremists to defeat centrists. The remnant of the median voter theorem in the incremental positioning model is a razor's edge. Nevertheless, there are at least some incentives for movement towards the center.

Thus, we begin to see a model of legislative extremism take shape, with some mysteries remaining. Strategically constructed legislative agendas can *reduce* the electoral damage of extreme positioning, but not *eliminate* it. Hence, even legislators with extreme preferences will need some extra incentives to construct ideologically extreme bundles of roll call votes, and that is where parties will come in.

Legislative Agendas and Electoral Preferences

The analysis so far has demonstrated that an electorate consisting of voters with single-peaked, symmetric preferences over *policy* will sometimes prefer the bundle of roll call votes required to adopt a relatively extreme position to the bundle of votes required to adopt a more centrist position, but that movement toward the median voter's ideal point will yield at least some electoral benefit. How dependent are these predictions on oddly structured legislative agendas? The electoral advantage of moving toward the center, and the electoral optimality of the median voter's ideal point, did *not* depend on an oddly constructed legislative agenda. It was the proximity paradox—the possibility of voters preferring distant candidates to proximate candidates—that relied on a legislative agenda with specific characteristics. In order for an electorate to prefer a candidate further from the median voter's ideal point to a candidate closer to the median's ideal point, the legislative agenda needed to include either asymmetric frequencies or asymmetric differentials (or both). Thus, in order to assess the degree to which the proximity paradox serves as a useful explanation for the electoral victories of ideologically extreme legislators, we must assess the occurrence of asymmetric frequencies and asymmetric differentials.

The political circumstances that produce asymmetric frequencies are intuitively easier to visualize in the context of theories of Congress, but as we shall see, they have more limitations as an explanation for the electoral success of extremists.

Beginning with asymmetric frequencies, consider the function of the committee system in the House of Representatives. Committees act, to some degree, as gatekeepers. If a bill is referred to a committee, and the committee either rejects or simply refuses to act on the bill, that bill will not receive a floor vote except in the highly unlikely event of a successful discharge petition.[1] Thus, one of the functions of the committee system in the House of Representatives is to prevent certain bills from reaching the floor. Furthermore, when the congressional reforms of the 1970s weakened the seniority system, the result was to give party leaders a degree of authority over this gatekeeping function. With that authority, they have the potential to block certain bills from receiving floor votes. Specifically, if two conditions are met, the majority party would prefer to block a floor vote: (*a*) the bill would move policy further from the majority party leadership's ideal point, and (*b*) the bill would pass on a floor vote. When these conditions are met, the majority party leadership would prefer to give the bill to a committee whose chair will refuse to schedule action, thus preventing floor consideration (and passage). According to the Cartel model of party influence in Congress, structuring the legislative agenda in this way is one of the primary methods by which parties operate (Cox & McCubbins 1993, 2005). If the majority party controls the agenda in this manner, the result is that the legislative agenda will contain asymmetric frequencies. Thus, it would seem that majority party agenda control can allow majority party extremists to defeat minority party centrists by creating legislative agendas in which voters prefer the bundles of votes offered by majority party candidates rather than the bundles offered by minority party candidates, even when the minority party candidates are more centrist.[2]

Consider the following. Suppose that the majority party leadership has a position of PL, and the floor median has a position of FM where PL < FM. Suppose, further, that an arbitrary district has a median voter whose ideal point is i_{MV}, and faces a choice between an incumbent located at MC and a challenger located at C where PL < MC < FM < i_{MV} < C.

Given this array of locations, what does Cartel theory suggest about the likelihood of asymmetric frequencies and the effect of those frequencies on the median voter's preference ordering between the incumbent's and challenger's respective bundles of roll call votes? Suppose that the majority party leadership is concerned not merely with policy outcomes, but with never losing a roll call vote. Then the majority party leadership will block a floor vote on any roll call vote, *j*, such that PL < $(sq_j + a_j) / 2$ < FM. Since PL < MC < FM < i_{MV}, the incumbent member of Congress will only cast a roll call vote contrary to the preferences of the median voter for a roll call vote, *j*, such that FM < $(sq_j + a_j) / 2 < i_{MV}$. The incumbent member of Congress is protected from having to cast any vote with a cut point in the (MC, FM) interval, which would have been electorally damaging, because party leadership blocks the vote. This kind of agenda control does nothing to help the challenger and, in fact, exacerbates the voters' disagreements with the minority party. This is in line with the Cartel model as an electorally motivated strategy of agenda manipulation.

However, this is something of a cheat in the context of *spatial* theory, and unfortunately this cheat was at the heart of the original derivation of the proximity paradox (Buchler 2011a). If the majority party leadership blocks a floor vote on any roll call vote with a cut point in the range of (PL, FM), then the majority party leadership, the incumbent member of Congress under examination, and the floor median will have substantively indistinguishable positions in the policy space. Functionally, then, the incumbent member of Congress cannot have, or *appear* to have, a policy position that is any further from the district median voter's preferences than the floor median. Thus it would be specious to argue that the legislative agenda has permitted a candidate further from the median voter's location to defeat a candidate closer to the median voter's location because empirically the incumbent member of Congress would appear more centrist than she actually is. The fact that we *observe* ideological extremists winning re-election suggests that we must look beyond this simplistic method of generating a proximity paradox with asymmetric frequencies.

Of course, there is no policy motive for the majority party leadership to block floor votes on bills that they want to pass if those bills cannot pass a floor vote. The preservation of the status quo is the result either way. Thus, the interval of (PL, FM) will not necessarily be devoid of *any* roll call votes. Instead, a policy-motivated majority party leader may permit a floor vote on a roll call vote, *j*, such that PL $< (sq_j + a_j) / 2 <$ FM if $sq_j > a_j$. Under these circumstances, the majority party leadership prefers the alternative to the status quo, but since the floor median prefers the status quo to the alternative, the bill fails. However, the policy result of permitting the roll call vote to occur is the same as the policy result from blocking the vote—the status quo. So the majority party leadership gains nothing by blocking it, except to save itself the embarrassment of having been rolled. That is a consideration, and majority party leaders do not like "getting rolled," but this is not an instrumental, policy-motivated strategic choice. Hence, we should not necessarily expect the interval of (PL, FM) to be devoid of *any* roll call vote cut points, and we should not expect the interval of (MC, i_{MV}) to be devoid of roll call cut points. Moreover, the existence of roll call vote cut points in these regions makes the locations of PL, MC, FM, i_{MV}, and C functionally and empirically distinct.

In contrast, though, let us consider the interval of (FM, C). There is no reason to expect *any* roll call vote, *j*, such that FM $< (sq_j + a_j) / 2 <$ C to be blocked from a floor vote. Hence, party influence on the legislative agenda provides *no* electoral protection for a position of C.

Combining these observations, what can we say about Cartel theory and the possibility that asymmetric frequencies induced by agenda control regularly allow extremists to defeat centrists? Agenda control under Cartel theory will reduce the number of roll call votes with cut points in a specific interval of the policy dimension. By definition, that means creating asymmetric frequencies. Moreover, those asymmetric frequencies can reduce the loss of utility the median voter would experience from the election of a majority party extremist. Thus, it can allow majority party extremists to defeat more centrist candidates of the minority party, as argued previously (Buchler 2011a). However, there is another important problem with posing asymmetric frequencies as an explanation for the electoral success of ideological extremists—it only protects members of the majority party. In fact, by the same token, the asymmetric frequencies created by negative agenda control should make it even *more* difficult for minority party extremists to win. Empirically, though, neither legislative caucus is a truly centrist caucus, and incumbents of both parties generally seem quite electorally safe, even if they overestimate their own electoral danger (Mann 1978). Extremists on both sides regularly win re-election, even though they vote on the same legislative agenda. Asymmetric frequencies alone cannot explain that. Asymmetric differentials, though, are more complicated.

Asymmetric differentials occur when the roll call votes with cut points in one region are more "important" than the roll call votes with cut points in another region. The complication is that "importance" is relative. What makes a roll call vote important to a voter is having a large gap between the utility of the status quo and the utility of the alternative. However, the magnitude of that differential varies by voter. Consider two voters. Voter 1 has a utility function for policy of $U_1(p) = -(6 - p)^2$, and Voter 2 has a utility function for policy of $U_2(p) = -(3.5 - p)^2$. Consider how these voters would respond to two hypothetical bills. Bill 1 would move tax policy from a location of 3 to a location of 4. Bill 2 would move immigration policy from a location of 3 to a location of 9. Voter 1 would receive −9 utils from the status quo on tax policy, and she would receive −4 utils from the alternative on tax policy. Thus, Bill 1 has a utility differential of 5 utils for Voter 1. Voter 1 is indifferent between the status quo and the alternative on Bill 2 (immigration policy) because the two options are equidistant from her ideal point. Each option yields her −9 utils, so the utility differential for Voter 1 on Bill 2 is 0. For Voter 1, Bill 1 is far more important than Bill 2.

However, the opposite is true for Voter 2. Voter 2 is indifferent between the status quo and the alternative on tax policy created by Bill 1 because the two options are equidistant from her ideal point of 3.5. Either way, she receives −.25 utils. On the other hand, while she would receive −.25 utils from the status quo on immigration policy, she would receive −30.25 utils from the alternative immigration policy that would be created by the passage of Bill 2. Thus, Bill 2 has a utility differential of 30 utils for Voter 2. Voter 1 cares more about Bill 1, but Voter 2 cares more about Bill 2. While asymmetric frequency is purely a characteristic of the legislative agenda, asymmetric differentials are in the eye of the beholder.

We can, however, generalize. Some bills propose to move policy further from the status quo than others. For any fixed distance between a voter's ideal point and a roll call vote cut point, the utility differential increases as the status quo and alternative points are further apart. Consider a voter with an ideal point of i evaluating a roll call vote with a cut point of $(sq + a) / 2 < i$, where $sq < a$. The voter in question prefers the alternative to the status quo, but by how much?

$$U(a) - U(sq) = -(i - \mathrm{a})^2 - -(i - sq)^2 = -(i - \mathrm{a})^2 + (i - sq)^2 = 2ai - a^2 - 2isq + sq^2 \quad (2.3)$$

Now suppose the voter evaluates another roll call vote with the same cut point, but policy alternatives further apart, where the status quo point is $sq - d$, and the alternative is $a + d$, where $d > 0$. The voter will prefer the alternative to the status quo in this new scenario as well, but with greater intensity. The voter's utility differential for this new bill is given by the following:

$$\begin{aligned} U(a+d)-U(sq-d) &= -(i-(a+d))^2 - -(i-(sq-d))^2 \\ &= -(i-(a+d))^2 + (i-(sq-d))^2 \qquad (2.4) \\ &= 4d(i-(a+sq)/2)+2ia-a^2-2isq+sq^2 \qquad (2.5) \end{aligned}$$

It is quite simple to show algebraically that the voter's utility differential is greater when the status quo and alternative points are further apart.

$$4d(i-(a+sq)/2)+2ia-a^2-2isq+sq^2 > 2ai-a^2-2isq+sq^2 \qquad (2.6)$$

$$4d(i-(a+sq)/2)>0 \qquad (2.7)$$

Inequality 2.7 must be true because, by assumption, $i > (a + sq) / 2$, and $d > 0$. Thus, when voters have quadratic loss utility functions, the further apart the status quo and alternative points are for any given fixed distance between a voter's ideal point and the roll call cut point, the greater the voter's utility differential between the status quo and alternative points. Thus asymmetric differentials occur for any given voter based on variation in the distance between a roll call cut point and the voter's ideal point (see Voter 1 on Bill 2, or Voter 2 on Bill 1), and based on the distance between the status quo and alternative points.

That latter possibility for an asymmetric differential has important implications for the legislative agenda. Some votes are objectively substantively small because they would not move policy very far from the status quo. For these votes, only voters whose ideal points are relatively far from the cut point will care very much, and even then, their preference intensity is only a function of risk aversion to policy alternatives far from their ideal points. However, there are other roll call votes that are objectively substantively big because they would move policy quite far from the status quo. For these votes, it is still true that the further one's ideal point is from the cut point, the more one will care about the vote, but only voters whose ideal points are *very* close to the cut point will be relatively indifferent. Hence, asymmetric differentials will occur any time there are substantively significant votes on the legislative agenda, with the effects magnified by variation in the locations of the bills' cut points.

The real question, then, is whether or not asymmetric differentials within a single legislative agenda can give electoral advantages to ideological extremists on *both* sides of the spectrum. After all, in any given congressional election cycle, ideological extremists in both parties win, and they had to cast votes on the same legislative agenda. So consider a legislative agenda consisting of seven roll call votes (Table 2.2).

The legislative agenda in Table 2.2 consists of seven bills across seven issues. Each bill moves policy to the left (again, we are assuming an agenda constructed by a party attempting to move policy in one consistent direction). Six of those changes are very modest changes to policy characterized by a proposal to move a very short distance from the current status quo within that policy area. However, Bill 4 is a major policy change, shifting healthcare policy from a far-right status quo to a far-left alternative. We will table the question of how much this exaggerates the legislative agenda of the 111th Congress, but substantively, this describes a legislative agenda with a number of small and essentially symbolic votes that allow us to distinguish empirically between degrees of extremism in voting records, and one major policy vote that divides the parties very cleanly around a "centrist" location of 4.5. Notice, also, that the legislative agenda is not characterized by asymmetric frequencies. Roll call cut points occur one unit apart in the range of [1.5, 7.5]. The primary distinguishing characteristic of this legislative agenda, then, is the fact that it consists of a set of relatively unimportant and essentially symbolic votes and one large, substantive policy change that cleanly divides the Right from the Left. It is worth noting that unlike the legislative agenda from Table 2.1, this legislative agenda requires that each policy area have a different status quo point, but that is a minor issue.[3]

The legislative agenda in Table 2.2 has the potential to benefit both liberal extremists and conservative extremists. Consider two districts. In District A, the

Table 2.2 **An agenda with symmetric frequencies and asymmetric differentials**

Bill	*Issue*	*Status quo*	*Alternative*	*Cut point*
1	Abortion	1.6	1.4	1.5
2	Student loans	2.6	2.4	2.5
3	Environment	3.6	3.4	3.5
4	Healthcare	8	1	4.5
5	Immigration	5.6	5.4	5.5
6	Taxes	6.6	6.4	6.5
7	Gun control	7.6	7.4	7.5

median voter has a quadratic loss utility function with an ideal point of 4. In District B, the median voter has a quadratic loss utility function with an ideal point of 5. Thus, District A leans *slightly* to the left, and District B leans *slightly* to the right.

Suppose that District A has an incumbent whose location in the policy dimension is 1, and a challenger whose location is 6. The district median voter's ideal point is closer to the challenger's location than to the incumbent's location. However, the legislative agenda creates a proximity paradox in which the median voter of District A prefers the bundle of votes associated with the more distant and more ideologically extreme incumbent to the bundle of votes associated with the more centrist and more proximate challenger. On each bill, the incumbent votes for the more liberal alternative, so the median voter's utility for the incumbent's bundle of roll call votes is as follows:

$$\begin{aligned} U_{MVA}(\text{incumbent}) = &-(4-1.4)^2-(4-2.4)^2 \\ &-(4-3.4)^2-(4-1)^2-(4-5.4)^2 \\ &-(4-6.4)^2-(4-7.4)^2=-37.96 \end{aligned} \tag{2.8}$$

On Bills 1 through 5, the challenger would have voted for the more conservative status quo, but on Bills 6 and 7, the challenger would have voted for the more liberal alternative. So the median voter's utility for the challenger's bundle of roll call votes is as follows:

$$\begin{aligned} U_{MVA}(\text{challenger}) = &-(4-1.6)^2-(4-2.6)^2-(4-3.6)^2 \\ &-(4-8)^2-(4-5.6)^2 \\ &-(4-6.4)^2-(4-7.4)^2=-43.76 \end{aligned} \tag{2.9}$$

So given this legislative agenda characterized by asymmetric differentials but not asymmetric frequency, the slightly left-leaning median voter in District A prefers the bundle of votes associated with the distant, ideologically extreme incumbent to the bundle of votes associated with the more proximate and more centrist challenger because $-37.96 > -43.76$.

What about District B, with the slightly right-leaning median voter whose ideal point is 5? Suppose that District B is represented by an extremely conservative incumbent located at 8, who is facing a less extreme challenger located at 3. The median voter in District B is closer to the somewhat liberal challenger than she is to the extremely conservative incumbent. However, that doesn't mean she prefers the bundle of votes offered by the challenger to the bundle of votes offered by the incumbent. The extremely conservative incumbent in

District B votes for the more conservative status quo on each of the seven bills, yielding the district median voter the following utility:

$$\begin{aligned} U_{MVB}(\text{incumbent}) = &-(5-1.6)^2-(5-2.6)^2-(5-3.6)^2 \\ &-(5-8)^2-(5-5.6)^2 \\ &-(5-6.6)^2-(5-7.6)^2=-37.96 \end{aligned} \quad (2.10)$$

The challenger in District B would have voted for the more liberal alternative on Bills 3 through 7, but she would have voted for the more conservative status quo on Bills 1 and 2. So the challenger's bundle of roll call votes would yield District B's median voter the following utility:

$$\begin{aligned} U_{MVB}(\text{challenger}) = &-(5-1.6)^2-(5-2.6)^2-(5-3.4)^2 \\ &-(5-1)^2-(5-5.4)^2 \\ &-(5-6.4)^2-(5-7.4)^2=-43.76 \end{aligned} \quad (2.11)$$

So the challenger's location is closer to District B's median voter than the incumbent's location, but the median voter prefers the bundle of votes associated with the distant and ideologically extreme incumbent to the bundle of votes associated with the more centrist and more proximate challenger because of another proximity paradox ($-43.76 < -37.96$).

Consider the strangeness of this split. The median voter in District A prefers a more distant *liberal* extremist to a more proximate *conservative*-leaning moderate because of a proximity paradox induced by the legislative agenda, and the median voter in District B prefers a more distant *conservative* extremist to a more proximate *liberal*-leaning moderate because of a proximity paradox running in the opposite direction induced by the *very same legislative agenda!* Moreover, this occurs with an agenda that does not mask incumbents' ideological extremism by saving them from having to cast votes that reveal their extremism. The legislative agenda achieves this outcome simply because the most substantively important vote exacerbates the cost of electing a representative who is across the central ideological divide at 4.5. When the substantively important roll call votes on the agenda include those that cleanly divide the left from the right, the structure of the legislative agenda can simultaneously help extremists on *both* sides of that divide.

In the case of the hypothetical agenda in Table 2.2, the vote that creates this effect is a healthcare vote. However, the vote on the 2010 Patient Protection and Affordable Care Act probably hurt Democrats more than it helped them (Nyhan et al. 2012), which will be discussed in more detail in chapters 3 and 6. However, the fact that more House Democrats defected from their party on the floor vote than Republicans suggests that the PPACA did not divide the party factions as cleanly as the healthcare vote in Table 2.2. More importantly, though, Democrats who lost after voting yes on the PPACA did not, predominantly, come from Democratic-leaning districts. Often, they came from Republican-leaning districts that the party picked up in the wave elections of 2006 and 2008. That is consistent with the principles of this demonstration. A Democratic incumbent who voted for the PPACA in a slightly Republican district will more closely resemble the disadvantaged challenger in District B than the advantaged incumbent in District A. The agenda in Table 2.2 will *hurt* an even slightly liberal-leaning moderate representing an even slightly conservative district, which is consistent with Nyhan et al.'s (2012) findings about the impact of the PPACA on 2010 legislative election results. More broadly, though, substantively important votes that *do* cleanly divide left from right can create proximity paradoxes that help extremists from both parties. While Cartel-style negative agenda control could only help majority party extremists, asymmetric differentials that occur under normal political circumstances can actually help extreme incumbents in both parties defeat centrist challengers at the same time, when the cut points for important votes divide the party factions cleanly along the central ideological break.

Thus, the incremental process by which legislators adopt their positions in the policy space can allow extremists of *either* party to defeat centrists of the opposing party, even when voters have single-peaked, symmetric preferences over *policy*, and pure policy motivations. Moreover, this can occur even when there are roll call votes that make it possible to distinguish empirically between legislators' locations throughout the entire policy space, rather than masking extremism by blocking the roll call votes that would have made it possible to see extremism. This is critical because our empirical observations about legislators' locations in the modern Congress through the NOMINATE procedure make distinctions between moderates and extremists that have predictive power. Therefore, it cannot be that extremists are protected from the electoral consequences of their extremism by legislative agendas that prevent them from having to reveal their extremism. Instead, voters are willing to elect and re-elect extremists despite the extremism of the voting records they build, and indeed, use to establish their locations in the policy space. As we have seen here, that can be rational, even when traditional spatial theory and basic intuition suggest that it shouldn't be rational.

Prospective versus Retrospective Voting

The model here is a retrospective voting model, so an obvious question is what would change if voters applied preferences over bundles of roll call votes to a prospective model? The basic calculus does not change—voters can prefer the bundles associated with distant candidates to the bundles associated with more proximate candidates. The key distinction is that the legislative agenda doesn't just change over time—it changes endogenously. If the voter's preference depends on the legislative agenda, and her vote helps determine which party sets the agenda, then she must consider not just which bundle of roll call votes she prefers, but which agenda. Returning to the more common sequence of election-then-policymaking should call our attention to the campaign dynamics of agenda setting. Democrats and Republicans do not merely announce their intention to *cast* different votes—they announce their intention to *call* for different votes. When Democrats announce their intention to push for votes on gun control—votes that would be blocked in a Republican-controlled House—they are suggesting the implications of a Democratic majority for the legislative agenda, just as Republicans signaled their future legislative agenda during the 111th Congress with promises to vote to "repeal Obamacare," which they began voting to do—symbolically if nothing else—when the 2010 election gave them a majority in the House, but not the Senate. Much of campaigning centers around a proposed legislative agenda, and the general approach used in this chapter to assess voters' preferences over bundles of votes given an agenda is equally applicable regardless of whether we are analyzing a past or future agenda. The purpose of this book is to draw attention to the electoral implications of legislative dynamics rather than campaign dynamics, but the mathematical lessons are equally important for both, and an important implication here is that there is need for a model focusing on the campaign dynamics of agenda setting when voters derive utility, not directly from winners' locations in the policy space, but from the bundles of roll call votes with which winners adopt their locations when the agenda is endogenous. That, however, must be the subject of future work.

Electoral Responses to Incremental Positioning

Bringing all of these observations together, we can now pose at least a partial answer to one of the central puzzles of the book. Why do extremists defeat centrists in legislative elections? Consider what the observations in this chapter suggest about the relationship between incumbent positioning and aggregate election results. Moving toward the ideological center of the district will increase an incumbent's vote share, but even when an incumbent is further from the median voter than the challenger, she will still frequently win. Moreover, a normal legislative agenda can produce this effect for ideologically extreme incumbents of both parties. All that is required is for there to be items on the agenda that move policy far from the status quo and divide the ideological factions cleanly around the central ideological break. When the agenda includes such items, liberal extremists can defeat conservative-leaning moderates in districts that lean at least slightly to the left, and conservative extremists can defeat liberal-leaning moderates in districts that lean at least slightly to the right.[4] While either type of extremist would increase her expected vote share by moving toward the center, her extreme voting record does not necessarily prevent her from defeating a challenger who is actually closer to the district median voter.

These predictions mesh well with empirical observations of legislative elections. Extremists can and do defeat centrists, but, comparatively, extreme incumbents tend to win lower vote shares than their more moderate colleagues (Ansolabehere et al. 2001; Ansolabehere & Jones 2010; Bovitz & Carson 2006; Canes-Wrone et al. 2002; Carson et al. 2010; Erikson & Wright 2000). The latter pattern made sense in the context of conventional spatial theory, but the former did not. However, as we have seen, *both* patterns are fully compatible with the model of electoral behavior derived in this chapter. Explaining these patterns merely requires understanding that legislative elections are choices between bundles of roll call votes, not choices between unitary policy outcomes. There are electoral rewards for centrism, but they are modest and mitigated by the legislative agenda. Put slightly differently, there are *punishments* for *extremism*, but they are modest and mitigated by the legislative agenda.

Two points must be made about these results. The first is that this chapter has not derived any equilibrium conditions. It has only demonstrated a set of possibilities by showing how a rational electorate would respond to hypothetical candidate locations and hypothetical legislative procedures. The questions that follow, then, are (1) how will strategic legislators position themselves in the policy space through the roll call voting process given these observations about rational voters? and (2) how will the legislative process be structured given the implications of that process for incumbents' positioning decisions and the electoral consequences of those decisions? These questions will motivate the analysis in chapters 3 and 4. As we shall see, the mediated consequences of ideological extremism will help us to explain incumbents' decisions to adopt extreme positions, and the legislative procedures that enable the process.

3

The Costs of Incremental Positioning

Chapter 2 demonstrated that when voters evaluate incumbents by the bundles of roll call votes they cast in order to adopt their locations in the policy space, extremists can sometimes defeat moderates. There is a reward for legislators who move toward the center, but it is far from determinative. This helps us explain the electoral success of extremists, but provides only a weak answer to the next question: why do incumbents adopt extreme locations in the first place? The fact that incumbents continue to be re-elected demonstrates, circularly, that they aren't *too* extreme to be re-elected, but why pass up the reward for moderation, small though it may be? The fact that regression models estimate positive effects for ideological centrism on incumbent vote shares means that many incumbents are not maximizing their vote shares. Why not?

Spatial theorists have posed two broad sets of answers. The first is that candidates must build support from more than just their general election constituencies. They must also build support from party activists (Aldrich 1983a, 1983b). Even more directly, they must first win primaries in order to get to the general election stage, and primary voters, like activists, are frequently noncentrists. Moreover, Brady et al. (2007) show that a candidate who is too centrist pays at least a modest electoral price during the primary. Thus, the structure of the electoral system does not put exclusively centrist pressure on candidates. However, we cannot rely on conflicting electoral pressures to explain polarization. The general election penalty for extremism is measurably greater than the primary election penalty for centrism, so it shouldn't be surprising that Abramowitz (2008) and Hirano et al. (2010) find that primaries have limited explanatory power with respect to polarization. Some basic historical observations should demonstrate why. A great deal of attention has been paid as of late to the fear that many members of Congress currently face, particularly in the Republican Party, of losing a primary to an ideologically extreme challenger if they are too moderate. As we shall see in chapter 5, though, these fears are not only overstated, suggesting that such a response would be irrational, but became particularly prominent *after* polarization had set in for the modern

Congress. The periodic emergence of primary threats described by Boatright (2013) does not match the steady trend toward polarization in the post–World War II House. Fear of primaries may contribute to extreme positioning in the modern Congress, and chapter 7 will describe how that has created unique intraparty conflicts within the Republican caucus since the 2010 election, but primary threats are far from a complete explanation for electorally suboptimal positioning.

The other common spatial explanation for legislators' willingness to adopt electorally suboptimal positions is that members of Congress simply have policy preferences themselves, with which they must balance their electoral ambitions. While there is appealing parsimony to Mayhew's (1974) model of legislators driven purely by electoral ambition, the more empirically realistic model of legislators' motivations comes from Fenno's (1973) observation that re-election is only one of three central motivations for members of Congress, along with policy interests and power and status within Congress. Based on the acknowledgment that candidates have policy preferences, a series of spatial models beginning with Wittman (1973, 1977, 1983) have analyzed the trade-offs between pursuing policy and re-election since the most electorally beneficial policy position is not necessarily a candidate's preferred policy.[1] These models provide an important starting point to explain legislators' willingness to adopt electorally suboptimal positions in the context of spatial theory.

The problem with existing Wittman-style models, though, is in the specification of the trade-off between policy and electoral security. In order to determine the degree to which strategic candidates will make policy concessions for *any* reason, electoral or otherwise, we must accurately calculate the cost of carrying out an insincere policy platform should the candidate in question win. The traditional method of doing so is to simply plug the location of that insincere platform into the candidate's utility function for policy. However, if the political process of carrying out a campaign platform is the incremental process of casting roll call votes, then candidates must calculate the policy cost of carrying out an insincere platform incrementally, just as voters must do when they calculate the utility of electing any given candidate.

The first point to be made about the utility that legislators receive from enacting a given policy platform, then, is that it is subject to the same paradoxes discussed in chapter 2. However, the relevant points of interest regarding legislators will be different for several reasons. First, candidates must balance their policy preferences with their electoral considerations, which is not true for voters. Second, centrist voters facing a choice between noncentrist candidates on either side of their ideal points must consider the relative costs of deviations in *two* directions, whereas candidates primarily face the question of how much to deviate from their preferences in *one* direction—toward the median voter.[2]

Thus, while the paradoxes of chapter 2 can apply to legislators, they are less relevant to legislative behavior. So this chapter will focus on another set of principles that follow from the incrementalism of legislative positioning.

If a member of Congress wants to position herself at any given location in the policy space, she must cast roll call votes as though that location is her true ideal point. Selecting an insincere location, then, means casting roll call votes inconsistently with one's personal policy preferences. Each time incumbents face roll calls on which their personal preferences conflict with their electoral interests, then, they must determine whether or not the marginal electoral benefit of an insincere vote outweighs the policy cost of that vote, and that process has been largely ignored by spatial theorists studying elections. Fowler and Frederking (2000) derive something analogous to a positioning model in which incumbents make policy decisions rather than simply choosing a policy location, and in an empirical model of roll call voting, Ladewig (2010) examines how legislators incrementally adapt their locations through roll call voting in order to respond to electoral considerations. However, these are the rare exceptions, and neither model traces out the effects of incremental positioning for spatial theory more broadly.

The problem with spatial theorists' failure to do so is that without accurately estimating the *cost* of selecting an insincere location, we cannot evaluate its *rationality*. Under most circumstances, it will be costless for legislators to adopt insincere locations in the policy space, even when those insincere locations are far from their ideal points, because individual legislators so rarely cast pivotal votes. Individually, then, every legislator has an incentive to adopt the location of her district median voter, regardless of her own policy preferences. Thus, once we understand the incremental nature of the positioning process, polarization in Congress becomes even more puzzling. However, when we examine the odd cases of legislators who *do* cast pivotal votes, we see that the policy costs of getting one's least-preferred outcome can be unboundedly large. That creates incentives to delegate power to party leaders, which will be the focus of the unified model in chapter 4.

Costless Positioning and the Importance of Being Pivotal

Chapter 2 built a model of voting behavior in which rational voters evaluate incumbents on both the pivotal and nonpivotal roll call votes they cast. The reasoning was twofold. First, since so few roll calls are decided by a single-vote margin, a voter who only cares about *pivotal* roll call votes will frequently be indifferent to an incumbent's voting record, making any form of modeling pointless. Second, retrospective analysis is intended as prospective guidance. An incumbent's nonpivotal roll call votes in the past provide an indication of how she will vote in the future, and such votes in the future *might* be pivotal.

Neither is a compelling reason to assume that *legislators* treat pivotal and nonpivotal roll call votes equally. First, legislators *do* have other considerations besides the direct policy effects of their roll call votes. Most importantly, they have re-election goals and power goals, as in Fenno's (1973) model. When legislators are not pivotal, they must simply base their decisions on their other goals, so even when legislators recognize that their votes won't determine policy outcomes because they aren't pivotal, there is still something for them to consider on a rational basis, and hence something for *us* to model.

Second, incumbents have no need to observe their own behavior to help them determine how they might behave in the future. Thus, while voters have compelling reasons to consider pivotal and nonpivotal roll call votes equally, incumbents don't. In fact, it makes no substantive sense to assume that they have policy preferences but *don't* consider the difference between pivotal and nonpivotal roll call votes.

This presents us with a problem. The policy cost of casting a nonpivotal roll call vote against one's own policy preferences is zero because, by definition, a nonpivotal vote does not affect the policy outcome. And, since members of Congress rarely cast a pivotal roll call vote on a consequential piece of legislation, regardless of how strong an incumbent's policy preferences are, they shouldn't matter. No matter how weak their electoral incentives are and no matter how strong their policy preferences are, cross-pressured legislators should simply defer to their constituents, as Groseclose and Milyo (2010) argue. Doing so means incumbents should incrementally adopt the locations of their district median voters by always voting in accordance with their median voters' preferences—recall from chapter 2 that the bundle of roll call votes associated with the median voter's location is a Condorcet-winning bundle, even in the incremental model. Incumbents *can* win when they are more extreme than their challengers, but if they can costlessly position themselves anywhere in the policy space, why take any risk? After all, as Mann (1978) argues, Members of Congress tend to overestimate the electoral dangers they face.

This should challenge our understanding of polarized voting scores in Congress. Empirical analysis consistently shows that extremism costs incumbents vote shares (Ansolabehere et al. 2001; Ansolabehere & Jones 2010; Bovitz & Carson 2006; Canes-Wrone et al. 2002; Carson et al. 2010; Erikson & Wright 2000). If it is electorally optimal for incumbents to adopt the positions of their district median voters, and they can do so costlessly because they do so incrementally (rarely if ever facing a pivotal vote), then why aren't they doing so? Why are they paying the costs that they are visibly, *measurably*, paying?

Electorally suboptimal positioning does, at least, appear to be sincere, if *perhaps* irrational. Consider retiring legislators, who no longer have any reason to make electoral concessions. Legislators' voting decisions *after* they decide to retire might be taken as their sincere, personal policy preferences unless they are somehow still subject to party pressure.[3] How do legislators behave after they decide to retire? For the most part, their roll call voting patterns after they decide to retire are quite similar to their voting patterns when they were still running for re-election (Lott 1987; Lott & Reed 1989; Lott & Davis 1992, Bender & Lott 1996; Rothenberg & Sanders 2000), although they may engage in more participatory shirking (Rothenberg & Sanders 2000). Members of Congress vote differently after their retirement decisions on no more than a few percent of roll call votes. This would seem to suggest that legislator's voting patterns are relatively sincere reflections of their personal ideal points, even when they are running for re-election. This occurs even though legislators receive lower vote shares for extremism (Ansolabehere et al. 2001; Bovitz & Carson 2006; Canes-Wrone et al. 2002; Carson et al. 2010; Erikson & Wright 2000), and yet have the capacity to increase their vote shares by voting as centrists, which they can do without policy cost most of the time because nonpivotal roll call votes have no policy cost. Why don't they?

What if legislators could force themselves to act, collectively, as though they were pivotal? How big might the policy *benefits* be? Chapter 4 will present a unified model in which an ideologically homogeneous but electorally diverse caucus empowers party leaders to enforce sincere voting so that the caucus can achieve policy victories that would otherwise be out of reach because in the absence of party discipline, everyone follows the logic described above. Critical to the decision to empower party leaders, then, is how appealing those victories are. We can make that determination by looking at what happens in the rare case that a legislator *does* cast the pivotal vote, because while rare, it happens.

When Legislators Are Pivotal

While legislators rarely cast pivotal roll call votes, it does happen, and sometimes on critical pieces of legislation like the 1993 Omnibus Budget Reconciliation Act and the 2010 Patient Protection and Affordable Care Act. So let us examine the hypothetical case of positioning for a pivotal legislator.

Suppose an incumbent has an ideal point, $i > 0$, and a quadratic loss utility function for policy: $U(p) = -(p - i)^2$. If i does not maximize her probability of re-election, she may instead locate herself at i^*, where $0 \leq i^* < i$. In conventional models, a policy-motivated candidate who wins the election on an insincere platform of i^* will receive a utility of $U(i^*) = -(i^* - i)^2 < 0$ instead of $U(i) = 0$. However, since actually carrying out a platform of i^* requires casting roll call votes as though i^* were her ideal point, that isn't the true policy cost of carrying out the insincere platform. On nonpivotal roll call votes, the cost of carrying out the platform is zero since the policy outcome is the same either way, but what about the case of the pivotal vote? Consider Figure 3.1.

Suppose the incumbent has electoral incentives to locate marginally to the left of i^*. At first glance, locating marginally to the left of i^* may seem relatively costless because i and i^* are close, and the incumbent's utility function is relatively flat in that region. However, that decision may be unboundedly costly. Consider a pivotal roll call vote between a liberal option, l, and a conservative option, c, where $(l + c) / 2 = i^*$. The legislator prefers c because $(l + c) / 2 < i$, but if she wants to locate marginally to the left of i^*, she must vote for l. The utility lost by doing so is not $U(i) - U(i^*)$, but instead $U(c) - U(l)$. The problem becomes clear when we consider another set of policy alternatives, $(l - z)$ and $(c + z)$, where $z > 0$. Anyone who prefers l to c will prefer $l - z$ to $c + z$ because $(l-z+c+z)/2=(l+c)/2$. However, it is simple to demonstrate that $U(c+z)-U(l-z)>U(c)-U(l)$. With respect to l and c, the legislator prefers c to l, and $U(c)-U(l)=-(i-c)^2--(i-l)^2=-(i-c)^2+(i-l)^2$.

$$=-i^2+2ci-c^2+i^2-2il+l^2 \tag{3.1}$$

$$=2ci-c^2-2il+l^2 \tag{3.2}$$

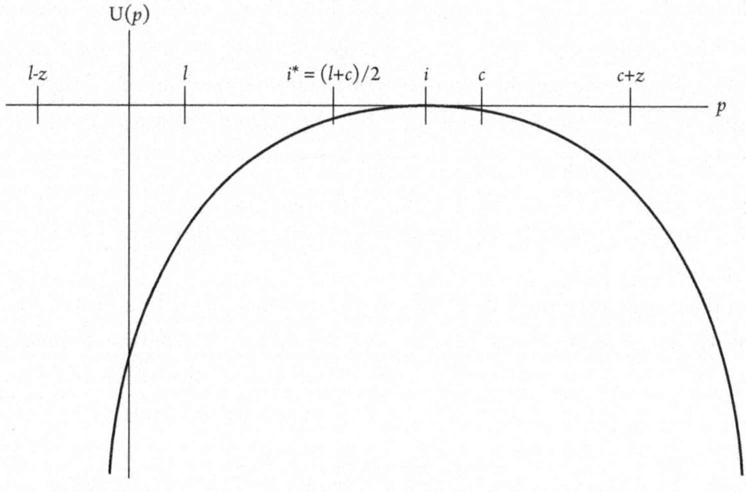

Figure 3.1 Alternative roll call votes with the same cut point

With respect to $l-z$ and $c+z$, the legislator similarly prefers $c+z$ to $l-z$. However, the utility differential between those alternatives is given by

$$U(c+z)-U(l-z)=-(i-(c+z))^2--(i-(l-z))^2$$
$$=-(i-(c+z))^2+(i-(l-z))^2.$$

$$=-\left(i^2-2i(c+z)+(c+z)^2\right)+\left(i^2-2i(l-z)+(l-z)^2\right) \quad (3.3)$$

$$=-i^2+2ic+2iz-c^2-2cz-z^2+i^2-2il+2iz+l^2-2lz+z^2 \quad (3.4)$$

$$=2ic+4iz-c^2-2cz-2il+l^2-2lz \quad (3.5)$$

It follows that $2ic+4iz-c^2-2cz-2il+l^2-2lz>2ci-c^2-2il+l^2$.

$$2ic+4iz-c^2-2cz-2il+l^2-2lz>2ci-c^2-2il+l^2 \quad (3.6)$$

$$4iz-2cz-2lz>0 \quad (3.7)$$

$$z(4i-2c-2l)>0 \quad (3.8)$$

$$z(2i-l-c)>0 \quad (3.9)$$

Inequality 3.9 must be true because, by assumption, $i>(l+c)/2$, and $z>0$. Thus, the cost of voting insincerely when faced with a choice between l and c is lower than the cost of voting insincerely when faced with a choice between $l-z$ and $c+z$, even though the distance between the legislator's ideal point and the cut point for each roll call is identical. Inequality 3.9 is vital to understanding the costs of strategic positioning and policy outcomes more broadly because it demonstrates that for any fixed distance between i and i^*, the cost of *voting as though* one's location is i^* may be unboundedly costly—it increases as a linear function of z. Higher values of z imply more substantively important bills because the status quo and alternative points are further apart. Note, also, that when the substantively important bills are those with cut points near the central ideological divide, that is precisely when chapter 2 showed that proximity paradoxes occur, protecting extreme incumbents in both parties.

Even if we impose a limit by restricting the policy space to a finite space, the maximum cost of an insincere vote is determined by a legislator's degree of risk aversion. Moreover, even if we eliminate risk aversion altogether *and* make the policy space finite, there is still no necessary upper bound on the *number* of roll call votes that must be cast with cut points in a given range, and if the cost of an insincere vote must be paid each time, it may still be unboundedly costly to adopt an insincere location in the policy space.

So, we return to Inequality 3.9. The cost of *voting as though* one is located at (or marginally to the left of) i^* rather than i may be unboundedly high, even if $|i - i^*|$ is arbitrarily small. Of course, that assumes a legislator is casting the pivotal vote, but abstracting away from the pivotal/nonpivotal distinction, the cost of sacrificing policy outcomes may be unboundedly large, which will become more directly important in chapter 4 as parties blur the distinction between pivotal and nonpivotal roll call votes, at the insistence of their members.

Pivotal Votes on Important Legislation

In order to understand how unboundedly large policy costs influence roll call voting, let us examine some empirical cases of legislators casting pivotal votes on major legislation. Consider the 1993 Omnibus Budget Reconciliation Act and the 2010 Patient Protection and Affordable Care Act. In the House, the 1993 budget was passed by a single vote, and in the Senate, the PPACA received the bare minimum number of votes for cloture. Moreover, these were particularly prominent votes, with electoral effects likely *much larger* than the average effect estimated by Canes-Wrone et al. (2002). Thus, they serve as valuable reference points for the possibility that the policy cost of voting insincerely may sometimes be too high to justify an insincere vote, even with abnormally *large* electoral consequences. More generally, they serve as case studies in how legislators may be willing to sacrifice electoral security for policy victories, even when the electoral costs are especially high and known in advance. Let us consider each case.

Bill Clinton's 1993 budget immediately brings to mind the name of a single representative—Marjorie Margolies-Mezvinsky (D-PA). Margolies-Mezvinsky represented a district that was far from solidly Democratic. In 1992, Bill Clinton won only 52% of the two-party vote in that district, and Margolies-Mezvinsky herself only won the 1992 election—her first—by a razor-thin margin of less than half a percentage point. Clinton's first budget was a controversial piece of legislation, particularly because it contained a set of tax increases. Up through the presidency of George H. W. Bush, the Republican Party was divided between a faction of "deficit hawks," who were willing to support tax increases in some circumstances to reduce the budget deficit, and "supply-siders," who were either uninterested in the budget deficit or believed that reducing tax rates would ultimately spur enough economic growth that even at a lower tax rate, total revenue would increase, thereby eventually reducing the deficit. However, after President George H. W. Bush agreed to a tax increase in violation of his "no new taxes" pledge from the 1988 campaign, the supply-side faction of the party gained clear control of the party apparatus, and the party oriented itself centrally around an objection to tax increases. So when President Bill Clinton proposed a budget in his first year that included tax increases, the budget became a flashpoint for partisan conflict. The Republican legislative delegation opposed the budget unanimously, but Democrats had a majority in both houses of Congress. They could win the vote, but only with minimal defections. The fate of the budget ultimately rested in the hands of Marjorie Margolies-Mezvinsky. The final vote on the conference report occurred on August 5, 1993. Without Margolies-Mezvinsky, there were 217 votes in favor of the conference report, including the vote from Speaker Tom Foley, and 216 against. Had she voted against the bill, the House would have been tied 217 to 217, which would have been a defeat, but she could cast the 218th vote necessary to pass the bill. Margolies-Mezvinsky was somewhat more moderate than the Democratic caucus at the time, but not far out of step. Her DW-NOMINATE score was –.1890, compared to a party median of –.310 in the 103rd House of Representatives. Nevertheless, the electorally optimal choice was clear—vote against the bill—and she had indicated that she was likely to vote no until the very end. However, she eventually decided to vote for the bill after long discussions with fellow Democrats and President Clinton. As she cast the vote, Republicans famously chanted, "Good-bye Marjorie," indicating their belief that the vote would cost Margolies-Mezvinsky her seat in the 1994 midterm election. As it turned out, Margolies-Mezvinsky lost her first and only re-election campaign in 1994 by about five percentage points in the two-party vote, along with 33 other Democratic incumbents. While it would be difficult to attribute her loss *solely* to that vote, popular legend does so, and it is at least plausible. Even though Canes-Wrone et al.'s (2002) estimates suggest that a single roll call vote generally has very modest electoral costs, there may still be individual roll call votes with much higher costs, and the 1993 budget may have been one of those votes.

So, did Margolies-Mezvinsky cast a sincere or an insincere roll call vote? After her defeat, Margolies-Mezvinsky expressed a great deal of pride in her supposedly career-ending vote. Prior to the vote, though, she had indicated opposition to the budget. This suggests three possibilities. First, perhaps her initial statements of skepticism were sincere, her ultimate vote was insincere, and her later statements of pride were merely lobbying efforts for a "profile in courage" award to save face after a plan gone wrong. Second, perhaps her initial statements of skepticism were merely electoral posturing. In that scenario, she wanted to vote against the budget for electoral reasons, but when faced with casting the *pivotal* vote, she decided that the marginal policy cost of defeating the budget was greater than the electoral cost of a sincere vote, even on such a high-profile, electorally salient bill. Finally, perhaps she just changed her mind on policy.

While we can never completely rule out the possibility of sincere conversion, the third option seems the least likely. Politically informed individuals have plenty of psychological mechanisms for rejecting arguments inconsistent with their existing beliefs, so actual conversion among the politically sophisticated is quite rare. That leaves two other possibilities—the initial statements of opposition were insincere, or the vote was insincere. If the expectation was that a yes vote would end her career—an expectation that House Republicans certainly had, given their chant—then it is difficult to see what would have convinced her to vote yes *other than* personal policy preferences. The vote was of central policy importance. Even if the cut point had been relatively close to Margolies-Mezvinsky's ideal point, the cost of receiving her least-preferred policy outcome was potentially unboundedly large by Inequality 3.9, and since she would cast the pivotal vote, the policy she received really was up to her. Moreover, the electoral stakes were known (or at least believed) to be significantly greater than Canes-Wrone et al.'s average cost estimate. Chapter 4 will address punishments and rewards for party loyalty on roll call voting, but given both the policy and electoral stakes, it is difficult to imagine what punishment or reward would outweigh the combined policy stakes and known electoral costs of a yes vote on the 1993 budget if a yes vote were pivotal and insincere.

If no conceivable carrot or stick could induce an insincere pivotal vote in favor of the 1993 budget, then the most plausible explanation is that Margolies-Mezvinsky initially expressed her intention to vote no, which would have been electorally optimal, because she intended to cast an electorally optimal vote. However, when faced with casting a *pivotal* vote, she chose the electorally suboptimal, but sincere, option of yes. In that interpretation, Margolies-Mezvinsky's eventual statements of pride in the vote reflected her actual policy preferences. That is precisely what the analysis above suggests should happen in this circumstance. The tale of Rep. Margolies-Mezvinsky suggests that even when the electoral costs of a sincere roll call vote are significantly *larger* than Canes-Wrone et al. (2002) estimate, the potentially unbounded policy cost of an insincere roll call vote may still make sincere voting the rational choice.

The complication with using Margolies-Mezvinsky as an exemplar of this logic, though, is that her statements over time were inconsistent with each other, and too much of the negotiation took place behind closed doors. The argument, then, relies on reading between the lines and attempting to determine which statements were sincere and insincere. While useful, then, the story should not be overemphasized. However, there is a much more clear example of the logic derived so far—healthcare reform in the 111th Senate. In the 111th Congress, undoubtedly the most significant piece of legislation considered was the Patient Protection and Affordable Care Act. The bill had three primary components. First, it would provide federal subsidies to help individuals purchase private health insurance if they could not afford it (or for lower-income individuals, expand eligibility for Medicaid). It would impose a set of regulations on health insurance companies to prevent them from denying coverage to individuals with "preexisting conditions." Finally, it would impose a tax penalty on those who could afford insurance, but chose not to purchase health insurance. For Democrats, the bill was an attempt to fulfill a long-standing central goal of expanding access to healthcare to those who had been unable to afford it. For Republicans, the bill represented an expansion of federal entitlements and regulation of economic choices. For each side, then, the bill was of critical importance.

In the 2008 election, Democrats expanded their majorities in the House and Senate and won the presidency. However, they only won 59 seats in the Senate, and even the seating of their 59th senator was delayed while Republicans contested Al Franken's victory over incumbent Senator Norm Coleman in Minnesota. Meanwhile, Republicans were committed to filibustering essentially any Democratic bill since the central premise of their position was that entitlements should not be expanded. Democrats stood no realistic chance of peeling off even a single Republican vote in the Senate for a dramatic entitlement expansion, despite Senator Max Baucus's negotiations with the bipartisan "Gang of Six" from the Finance Committee.

Hence, the opportunity for Democrats to pass a healthcare reform bill was created only by Senator Arlen Specter's decision to switch from the Republican Party to the Democratic Party. Specter switched parties because he was about to face a rematch for his Senate seat in Pennsylvania. Specter was one of the few moderates in the Senate, and in 2004, he narrowly defeated a conservative challenger, Representative Pat Toomey, in the Republican primary. In 2010, he was likely to face Toomey again, and he did not think he could win. So he switched to the Democratic Party, believing that he would have a better chance at the Democratic nomination than at the Republican nomination while the Republican Party underwent a supposed purge of moderates from its ranks. In fact, Specter lost the Democratic nomination to Representative Joe Sestak anyway, so the tactic didn't work. However, in order to secure official party support in the Democratic primary, Specter engaged in some negotiation with Democratic leaders at the time. The precise contents of that negotiation are not public, but one must assume that Democratic leaders demanded Specter's promise not to join a Republican filibuster of a healthcare reform bill. With Specter in the Democratic caucus, the party held 60 seats—just enough to invoke cloture on a healthcare reform bill as long as *every* Democrat voted for cloture.

The trick is that when there are precisely 60 votes for cloture, every senator voting for cloture is casting a pivotal vote. A single defector would allow the Republicans to sustain their filibuster, thereby blocking the bill. This made the legislative negotiation difficult. Several Democrats, most visibly Senators Ben Nelson of Nebraska, Mary Landrieu of Louisiana, and Joe Lieberman of Connecticut made a series of demands regarding the content of the bill, and if the Democrats were to invoke cloture, they needed the support of each of these legislators. This allowed the wavering Democrats to make essentially any demand they could imagine, knowing that the demand would probably be fulfilled. Nelson, Landrieu, and Lieberman played their hands much more aggressively than any representative during the 1993 budget vote, perhaps demonstrating the relative strength and autonomy of senators and representatives. In the end, though, it came down to a cloture vote, and each Democratic senator cast a pivotal vote.

The electoral dangers were clear. The bill was not especially popular. Slim but consistent majorities opposed passage of the bill in every poll, but the opponents of the bill felt more strongly about the bill than the supporters. For Democrats representing Republican-leaning states like Nebraska and Louisiana, as Nelson and Landrieu did, the danger was even more clear. Eventually, empirical studies demonstrated that Democrats who voted for the PPACA fared worse in their 2010 re-election bids than Democrats who opposed it, controlling for district partisan leanings (Nyhan et al. 2012), and most Democrats expected this to be the case going into the vote. Moreover, Majority Leader Harry Reid had virtually no leverage over wavering Democrats. Chapter 4 will present a unified model in which party discipline becomes the key to positioning, but that model is primarily intended to explain the process in the House. This story takes place in the Senate. Consider Senator Joe Lieberman. Lieberman had been a consistent supporter of the Iraq war, and in 2006, that support cost him the Democratic nomination to retain his Senate seat. However, rather than accept defeat, Lieberman announced his intention to run as an independent candidate, against the official Democratic nominee, Ned Lamont. Lieberman won and was accepted back into the caucus. In 2008, he took that a step further and actively campaigned with and on behalf of the Republican nominee for president, Senator John McCain (R-AZ). By this point, Reid was under some pressure to at least deny Lieberman the chair of the Homeland Security Committee. Reid did not. He allowed Lieberman to keep his position as a committee chair. If Reid was not going to retaliate against Lieberman for these actions, the wavering Democrats knew that Reid would not impose any sanctions on any Democrat for anything.

When facing the final cloture vote, then, wavering Democrats had a choice. If even one of them defected, the bill would fail, and the Democrats would have lost their best chance at passing a healthcare reform bill in decades. However, voting for the bill entailed accepting an electoral penalty, and the chamber leader, Sen. Harry Reid (D-NV) had no ability to put pressure on wavering legislators (or at least no credibility if he wanted to make any threats). However, while wavering Democrats negotiated aggressively for a set of policy concessions, all had expressed support for the main provisions of the bill, and there is no reason to think that these statements were insincere.[4] This is precisely the calculation derived so far in this chapter, and an even clearer case than Margolies-Mezvinsky because she had made statements indicating more direct opposition to the Omnibus Budget Reconciliation Act of 1993 before ultimately voting for it.

So Democratic senators faced a clear strategic choice. Every one of them was in the position of casting a pivotal vote for cloture on a bill that they all wanted to pass, but where the electorally optimal vote would be against cloture. They could either take a policy victory in exchange for adopting an incrementally more extreme, electorally suboptimal position in the policy space, or they could sacrifice policy to adopt incrementally more centrist positions. They faced no credible threat of punishment within the chamber. This was a simple choice between the marginal electoral costs and marginal policy costs. In the end, all Democratic senators cast the electorally risky vote in favor of cloture on the PPACA. In doing so, they cast pivotal votes for policy victories because the cost of a policy defeat on such a central issue was just too high. Nelson and Lieberman didn't bother seeking re-election in 2012,[5] and Landrieu lost her 2014 re-election campaign. But they had the policy victory.

In the midst of legislative wrangling over the PPACA, former-Representative Marjorie Margolies-Mezvinsky herself published an op-ed in the *Washington Post* on March 18, 2010, comparing nervous Democrats to her situation in 1993. She wrote, "I am your worst-case scenario. And I'd do it all again." She was urging Democrats to cast electorally risky, and perhaps even electorally terminal, votes because the policy cost of defeating the PPACA would be worse than the electoral costs of supporting it.

During the 2010 midterm elections, 52 Democratic incumbents lost their House re-election campaigns. Many of the losing incumbents had defected from the party on the PPACA, and were defeated because they simply represented districts that the Democratic Party could not hold in a partisan environment like 2010, with unified Democratic control of the House, Senate, and presidency, along with a weak economy. However, empirical analysis has suggested that some of the losses among Democrats who had voted in favor of the PPACA were because of those very roll call votes (Nyhan et al. 2012). Like the 1993 budget, the PPACA may be that rare example of a roll call vote with potentially large electoral consequences that dwarf the average costs estimated by Canes-Wrone et al. (2002). Yet even on these two high-profile, electorally salient roll call votes, when legislators were put in a position of having to cast a pivotal roll call vote, they chose the electorally dangerous vote rather than the electorally safe vote. Chapter 4 will present a model in which legislators cast electorally dangerous votes because of party discipline. However, party discipline seems unlikely in these cases. For Representative Margolies-Mezvinsky, the clear and predictable electoral consequences (recall the chanting as she cast her vote) make it difficult to imagine what form of party discipline could have swayed her, but it was at least possible that party disciplinary threats played a role. For senators facing a cloture vote on the PPACA, though, party discipline is completely implausible. Reid simply had no credible threat to impose any punishment for disloyalty. Instead, then, the best explanation for Margolies-Mezvinsky's 1993 vote, and even more so for Landrieu's, Lieberman's, and Nelson's 2010 votes, is as follows: when a legislator is pivotal, voting insincerely on major legislation is too costly, even in the face of predictably harsh electoral punishments. For senators in 2010, this was even true with the memory of Representative Marjorie Margolies-Mezvinsky never far from the surface.

When legislators expect to cast a pivotal vote on a bill of central importance, the policy cost of an insincere roll call vote may be unboundedly high. That may prevent legislators from casting electorally optimal votes, thereby explaining some part of why legislators incrementally adopt positions in the policy space that appear ideologically extreme and electorally suboptimal. Moreover, this can be true even when the electoral cost of a sincere roll call vote is significantly higher than average.

Of course, this only works in the rare case of a pivotal vote, but the underlying calculation still matters. The question is, *if a legislator could* achieve policy victories such as those described here, for an electoral cost, would she? If her policy preferences were sufficiently strong, then she would. But how? Parties. If legislators can empower party leaders to enforce sincere voting, then they can achieve potentially unboundedly large policy *gains* that would be inaccessible otherwise because of Groseclose and Milyo's (2010) logic. That will be the thrust of the unified model in chapter 4.

The True Costs of Strategic Positioning

Chapter 2 showed how rational voters might re-elect ideologically extreme incumbents, even when presented with more moderate challengers, which can explain why extremists manage to win re-election with merely diminished margins. However, taking the observations from chapter 2 together with chapter 3, we have yet to explain why incumbents adopt extreme locations in the first place. Even though rational, centrist voters can sometimes prefer extremists to relative centrists, they will reward *movement toward* the ideological center with higher vote shares. Moreover, recall from chapter 2 that the bundle of roll call votes associated with the district median voter's position is a Condorcet-winning bundle. Incumbents *can* win with noncentrist positions. However, if incumbents can incrementally adopt their district median voters' positions without cost because they are so rarely pivotal, which this chapter argues, then why not do so? Why take even the slightest risk? After all, Mann (1978) argues that Members of Congress *over*estimate the electoral dangers they face. So, if they can costlessly position themselves anywhere in the policy space because the process is incremental, why are incumbents adopting electorally suboptimal positions?

In the rare case of casting a *pivotal* roll call vote, the policy cost that an incumbent pays for an insincere vote can be unboundedly large, making sincere voting rational, even when the electoral costs of sincere voting are also unusually high. Being pivotal is simply rare. But what if legislators could find a way to insert their own policy preferences into the process when they are not pivotal? If they could, the policy benefits may be unboundedly large, and that sets a direction for the unified model in chapter 4.

The next chapter will combine the observations from this and the preceding chapter into a unified model of legislative elections, parties, and roll call voting in order to explain the remaining puzzles about polarization—why do incumbents adopt ideologically extreme, electorally suboptimal positions, and what role do parties play in the process? As we shall see, parties become the mechanism by which ideologically extreme members of Congress force themselves to build electorally suboptimal voting records. For an ideologically homogeneous but electorally diverse caucus, sincere voting is a collective action problem. When policy-motivated, though, such a caucus will empower party leaders to use disciplinary measures to enforce sincere voting. By empowering party leaders in this manner, the caucus will achieve policy victories, but at an electoral cost because the consequence of that party discipline is to force caucus members to incrementally adopt ideologically extreme, electorally suboptimal positions in the policy space. However, those electoral consequences are themselves mitigated by the electoral paradoxes derived in chapter 2.

4

A Unified Spatial Model of Congress

The previous chapters provide the necessary components for a unified spatial theory of Congress, but on their own, neither served as compelling explanations for legislative polarization. Chapter 2 demonstrated that an ideologically extreme incumbent can defeat a more moderate challenger, even when voters are fully informed policy seekers with single-peaked, symmetric preferences. Even in that model, though, the median voter's location remains electorally optimal, and as discussed in chapter 3, unless a member of Congress is the pivotal legislator, her policy preferences shouldn't influence her roll call votes at all. She should simply defer to her constituents, meaning her district median voter, and incrementally adopt that location. What is needed to explain polarization, then, is an actor capable of inserting legislators' policy preferences into the decision calculus when they wouldn't otherwise matter: parties. Parties can create a disciplinary system in Congress, yet party leaders are elected, so any disciplinary system must reflect the desire of the caucus for that system. If caucus members place sufficient weight on policy goals, then, they can empower party leaders to enforce sincere voting, thereby inserting their policy preferences into roll call voting regardless of whether or not individual members are pivotal, and as long as the agenda meets the minimal conditions discussed in chapter 2, legislators will be insulated from the electoral damage done by the disciplinary system for which they vote. So the unified spatial model will center around a caucus vote for a leadership position and then trace out the effects of that leadership contest on the roll call voting process and a subsequent round of legislative elections in order to explain why members of Congress incrementally adopt ideologically extreme, electorally suboptimal positions, and why constituents re-elect them anyway.

The idea that members of Congress would *want* to be pressured to vote a certain way on pieces of legislation will strike many as odd, not just because people generally don't want constraints on their choices, but because discussion of party discipline is generally tainted by an implicit assumption that party pressure means pressuring legislators to vote in ways that violate their personal policy preferences. However, the model in this chapter will be based around an alternative conception of party influence that will be called "preference-preserving influence." This form of influence will be contrasted with "preference-deviating influence." When parties apply preference-deviating influence, they pressure members of Congress to vote in ways that conflict with the members' personal beliefs about which policies are good for the country. For example, if the Democratic Party were to pressure a moderate legislator who personally believes that tax increases weaken the economy to vote for a tax increase, that would be preference-deviating influence. However, when parties apply preference-*preserving* influence, they prod cross-pressured legislators to *adhere* to their sincere policy preferences rather than to make electoral concessions. For example, if the Democratic Party pressures a legislator who believes that the Patient Protection and Affordable Care Act is good policy to ignore the polls showing public opposition and vote yes, that is preference-preserving influence. The model in this chapter will derive the conditions under which the caucus empowers party leaders to apply preference-preserving influence, and trace out the effects of this influence on the process by which legislators incrementally adopt their positions in the policy space, and the electoral implications, thereby linking spatial models of elections to spatial models of the legislative process through the mechanism of party leadership.

Thus, the chapter will go beyond simply linking spatial models of elections to spatial models of the legislative process. An equally important component of the argument will be a reformulation of the "conditions" for "party government." Under this model, the conditions under which a legislative caucus empowers party leaders to impose discipline will be (1) ideological homogeneity, (2) electoral *hetero*geneity, and (3) strong policy preferences. Condition 2 will be the most important break from conventional CPG. The caucus with the strongest interest in party discipline will be one in which legislators have similar personal policy beliefs but represent different types of districts and hence face different electoral pressures. These reframed conditions for party government will absolve CPG of the charge that it borders on tautology. As we shall see, an ideologically homogeneous but electorally diverse caucus will *not* vote cohesively unless its members empower party leaders to apply preference-preserving influence, because in the absence of that influence, caucus members will remain responsive to their diverse electoral pressures. It is only through party influence that they will express their homogeneous personal preferences in the roll call voting process, thereby adopting extreme positions in the policy space.

Motivation: The Collective Action Problem of Sincere Voting

Sincere voting in Congress is subject to a collective action problem. That problem comes from a conflict that some members face between their personal beliefs about what constitutes good public policy and their desire to be re-elected. For a given roll call vote, suppose that some members of Congress have personal preferences that conflict with their constituents' preferences. This would mean that the midpoint between the status quo and alternative points for the bill lies between these legislators' ideal points and their district median voters' ideal points. In the absence of party influence, how should such members of Congress vote?

In a model without parties, Groseclose and Milyo (2010) address a nearly identical question about what happens when legislators face a conflict between their "position-taking" preferences and their "outcome" preferences. In their model, the conflict occurs when legislators face something like a poison pill amendment. Suppose that a legislator prefers an amended bill to an unamended bill, and an unamended bill to the status quo, as do her constituents. However, suppose that the unamended bill can pass a floor vote, but the amended bill cannot. Furthermore, suppose that the legislator's constituents do not understand the concept of a poison pill amendment. Absent any constraints, a rational legislator would vote against the amendment because a vote for the amendment is functionally equivalent to a vote for the status quo since the amended bill would be defeated on a floor vote. In contrast, a vote against the amendment is functionally equivalent to a vote for the unamended bill rather than the status quo. Since she prefers the unamended bill to the status quo, in the absence of electoral constraint, she would vote against the amendment. Groseclose and Milyo classify doing so as voting consistently with one's outcome preference. However, the legislator's constituents may not understand that voting against the amendment actually produces an outcome that they would prefer, so they will punish the legislator unless she votes in favor of the amendment. Groseclose and Milyo describe voting for the amendment as voting with one's position-taking preference. The equilibrium prediction is that regardless of how strong their outcome preferences are, and regardless of how weak the electoral pressure is, *all* legislators should vote consistently with their position-taking preferences.[1] While surprising, this result is intuitively simple. Unless a cross-pressured legislator expects to cast the pivotal roll call vote, which is rare, her vote will not affect the policy outcome. So voting consistently with her outcome preferences entails paying an electoral cost without any policy benefit. Regardless of the strength of her outcome preferences and regardless of how weak the electoral pressure is, she has no incentive to vote against the poison pill amendment. Only in the rare circumstance that a legislator is pivotal should her outcome preferences matter at all. Hence, when members of Congress face electoral pressure to vote in ways that conflict with their outcome preferences, the solution is for all members to ignore their outcome preferences and vote the way their constituents want.

Groseclose and Milyo assumed that legislators and their constituents shared the same outcome preferences, but that voters had a less sophisticated understanding of the legislative process. However, the logic of their model is directly relevant to the question for this chapter. What happens when some legislators have personal preferences over policy outcomes that conflict with the outcome preferences of their constituents? Suppose that a legislator personally prefers for a bill to pass, but her constituents prefer the status quo. How should she vote? This chapter will refer to a legislator's decision to vote for her personally preferred policy outcome as a "sincere" vote, and it will refer to a legislator's decision to vote for her constituents' preferred policy outcome as an "insincere" vote.[2] Following the logic of Groseclose and Milyo, the equilibrium result is for all legislators to vote consistently with their constituents' preferences, regardless of whether or not that means voting insincerely, so long as this does not result in a final vote margin of a single vote. After all, if nobody is casting a pivotal vote, then there is no reason for anyone to pay an electoral cost for voting sincerely, regardless of how strong their policy preferences are, and regardless of how weak the electoral pressure is. This effect can cause a bill to fail even when a majority in the legislature would sincerely prefer for the bill to pass. Hence, sincere voting as defined here may be a collective action problem by Olson's (1966) criteria. Passage of a bill is a collective good, and no single supporter determines whether or not that good is provided because nobody is pivotal. Thus, any supporter who might incur an electoral cost by voting for the bill will "free-ride" by voting insincerely, and, collectively, such free-riding causes the bill to fail.

In the absence of party influence, then, legislators' policy preferences shouldn't matter, and members should incrementally adopt the locations of their district median voters since even though they can sometimes beat more centrist challengers, there is nothing to be gained by not maximizing their vote shares, which they do by voting as though their district medians' positions are their own. If sincere voting is a collective action problem, then explaining polarization requires finding a solution to that collective action problem.

Enter parties. One of the most important functions that parties serve in Congress is to solve collective action problems, as Aldrich (1995), Cox and McCubbins (1993, 2005), and many others have argued. This particular collective action problem cries out for parties. Volden and Bergman (2006) argue that party strength is a strategic choice by its members, and building on that, Patty (2008) derives a model in which a majority legislative caucus empowers its party leader to impose discipline in order to solve a policymaking collective action problem very similar to the one posed above. In Patty's model, a legislative caucus strategically selects the amount of strength that the party leader will have in order to enforce discipline based on the amount of discipline that will be necessary to solve the collective action problem. Patty's focus is on the relationship between caucus size and equilibrium party strength. As we would expect intuitively, he argues that a smaller majority caucus requires more party strength because the number of defections that the caucus can accept and still win is determined by its size. However, Patty's approach has the potential to explain more about Congress than merely the relationship between caucus size and party strength, and it will be central to the structure of this model.

The unified spatial model of legislative elections, parties, and roll call voting will derive the conditions under which a legislative caucus selects a party leader who solves the collective action problem of sincere voting, reframing those conditions as the conditions for party government in Congress. Party influence will not exist independently of electoral influence; it will counteract electoral influence because if party influence does *not* counteract electoral influence, then even policy-motivated legislators should behave as pure re-election seekers by adopting centrist positions, even though doing so will result in their collective failure to achieve policy goals.

Other models that have used party pressure to explain polarization, such as Theriault (2008) and Van Houweling (2012), have focused on the ways in which party influence on procedure and procedural votes rather than substantive votes affect polarization in voting scores. However, this model will focus on party pressure on substantive votes. Moreover, while parties may have incentives to structure procedure in a way that allows caucus members to conceal some amount of their extremism, the fact that we observe polarized voting scores means that we must explain, as this model will do, how party pressure can operate when it fails to conceal extremism. Nothing in this model is intended to deny the importance of party pressure on procedural votes, but merely to show the additional importance of pressure on substantive votes.

Model Structure

Because the model in this chapter will link so many disparate models, it must be structurally unusual. The most unusual feature will be the order of events. Traditional spatial models of elections begin with candidates selecting their positions in the policy space. Then an election occurs. Finally, the winner implements her platform, generally by providing all voters with a policy equivalent to her platform position. The unified model here will reverse that traditional order, thereby focusing our attention on a very different type of strategic interaction.

By placing elections first, traditional models look for explanations about candidates' positioning decisions in the strategic interaction between competing candidates within the same electoral units. This has the unfortunate consequence of drawing attention *away* from the strategic interaction that occurs within Congress. The process by which legislators actually adopt their locations in the policy space is the incremental process of casting roll call votes, based on which we infer their locations. The structure of that process is determined by the strategic interaction between legislators across different electoral contexts, and it is that neglected process that the model here will address.

So, recall the critical point that every election except 1788 was also preceded by a complete legislative session. Unless we intend to capture the complete, two-century history of Congress with one model, then, there is no necessary *structural* reason to place the election first, and hence no necessary reason to focus on the interaction between incumbent and challenger rather than the interaction between different incumbents across districts. Moreover, most congressional elections are contests between an incumbent and a challenger. Through the roll call voting process, those incumbents incrementally adopt positions that conventional spatial theory suggests should be too extreme to win. Yet they win. Perhaps, then, we should place the legislative session first in order to explain the strategic interaction within Congress that produces extreme incumbent positioning through the roll call voting process, and ask whether or not that very same process can explain the electoral consequences of extreme positioning.

That is not to say that spatial models of elections *must* place the legislative session first. Legislators interact strategically with each other and with their respective challengers. Hence, neither model focuses on a more "real" process, and as we shall see, each model structure has benefits and drawbacks. Rather, the point is that a complete spatial understanding of Congress and legislative elections requires considering both types of models because they address different, but equally important, questions.

So the structure of the model will consist of three stages. The first stage will be a leadership contest within the majority party legislative caucus. The focus will be on the majority caucus so that the agenda does not block floor votes on bills supported by both rank-and-file and leadership, but the basic reasoning will apply equally to the minority caucus. Two legislators who aspire to lead the majority party caucus, and hence the chamber, will announce their platforms. Their platforms will consist of a disciplinary system describing the circumstances under which the party leader would impose punishments on a given rank-and-file legislator, and the magnitude of the punishment for any specific offense. After announcing their platforms, the caucus will vote for a party leader, who will also become the chamber leader. Then, in the next stage of the model, the chamber leader will carry out her platform. The result will be a sequence of roll call votes on which legislators will balance their policy preferences, the predicted electoral consequences of any given vote, and any relevant threat of punishment from the party leader in order to determine how they will vote. Through that sequence of roll call votes, each legislator incrementally adopts a position in the policy space with a bundle of roll call votes for her constituency. Then, in the final stage of the model, the elections will occur in which voters in each district respond to the votes that their incumbents have cast as referenda on the incumbents' performances.

The model will be unusual in another sense because it will abstract away from the behavior of challengers. Since positioning occurs as the incremental process of casting roll call votes and nonincumbents, by definition, have not cast roll call votes, voters have a problem. In principle, they could assess a challenger's policy location based on campaign statements and impute a bundle of roll call votes given that location, but there is a simpler and more realistic approach. Talk is cheap, but a challenger's party affiliation conveys a lot of information, particularly in the modern era of relatively homogeneous parties. Party affiliation is a powerful signal to voters about policy in the absence of other information (Snyder & Ting 2002), and adherence to party positions can provide candidates with electoral benefits anyway (Sniderman & Stiglitz 2012). Party reputation, then, serves as a more meaningful cue about a challenger's likely roll call voting patterns in office than the more complicated process of extrapolating individual roll call votes from campaign rhetoric. A voter making inferences, then, can impute a bundle of roll call votes for the challenger based on how representatives who are members of the challenger's party tended to vote. In an era of homogeneous parties, the voter may simply assume that the challenger would have voted the way her party median *did* vote, or how her regional median did vote (i.e., assume that a southern Democratic challenger in 1960 would vote as the median southern Democrat actually did vote). Challengers don't cast roll call votes, so they don't establish locations the same way as incumbents. However, they are constrained by party reputation, and the model should reflect that.

Assumptions

1. *Actors*. The actors consist of (i) two leadership candidates, (ii) the rank-and-file members of the party caucus that they are competing to lead, and (iii) each legislator's district median voter.

2. *Legislators' preferences.* Each legislator, *j*, has a quadratic loss utility function for policy within a single policy dimension, $U_j (p) = -(p - i_{MCj})^2$, where i_{MCj} is Member *j*'s sincere ideal point reflecting her personal beliefs about which policies are best. Each legislator is also concerned with re-election, and receives $V > 0$ independently of policy if she is victorious in her next re-election. Without loss of generality, we will examine the case of a right-leaning majority caucus in which legislators *tend* to have ideal points to the right of the minority party.
3. *District median voters.* Each district has a median voter with a quadratic loss utility function for policy, $U_{MVj} (p) = -(p - i_{MVj})^2$. Chapters 5 and 7 will address whether or not legislators have shifted from worrying about general elections to worrying about primaries, but for the purposes of this chapter, we will examine general election medians, in which case the direction of electoral pressure is more likely to be toward centrism rather than toward extremism.

The electorally optimal roll call vote is always to vote consistently with the preferences of the district median, and in the case of the general election median, this will tend to mean voting in a more centrist manner than legislators' personal preferences. If, for example, $i_{MVj} < i_{MCj}$, then the median voter and the legislator will only conflict on a roll call vote if the cut point places the voter on the same side of that cut point as the challenger's *party*. If the legislator "shirks" her representational duties and votes according to her policy preferences, the challenger wouldn't have done so. The median voter should be displeased. However, an incumbent *can* defeat even a more centrist challenger, as we learned in chapter 2. There is pressure on the member of Congress to vote in a manner consistent with the preferences of the district median, but not determinative pressure, except in the unlikely hypothetical case of facing an opponent located *at* the median voter, which challengers cannot actually do because they are constrained by party reputation anyway.

In order to generalize, then, this chapter will assume that a single electorally suboptimal roll call vote reduces the incumbent's probability of re-election by some increment.

This approach avoids the problem of the knife-edge equilibrium. If we attach a specific utility value to the median voter's assessment of the incumbent's bundle of roll call votes and compare that to the challenger's imputed bundle, as in chapter 2, then the incumbent will be content to cast unpopular roll call votes even without party pressure until she reaches the 50% threshold in the upcoming election. Given incumbents' electoral risk-aversion described by Mann (1978), this seems unreasonable. Rather, the model will simply incorporate incremental losses in re-election probability, recognizing that elections always operate under some uncertainty.

The question, then, is the magnitude of the effect. We can infer a magnitude from existing empirical analysis. Canes-Wrone et al. (2002) regress incumbent vote shares on ADA scores, and find that every 25 points of extremism cost an incumbent between one and three percentage points in an election. With 20 ADA votes per year, that translates to losing at most 0.6 percentage points per ADA vote. While that is small, the actual effect of most votes will be even smaller. If ADA votes are a sample representing a legislator's overall voting pattern, then each ADA vote does not put 0.6 percentage points at stake. Instead, all votes *represented by* a single ADA vote *collectively* put 0.6 percentage points at stake, and each individual vote is worth considerably less. Alternatively, if those effects really do come from the specific ADA votes themselves rather than the other votes that are correlated with ADA scores, then any non-ADA vote is truly inconsequential. In either case, a single roll call vote *usually* has very little effect on an incumbent's probability of re-election. Hence, we will assume that any one unpopular roll call vote will have an arbitrarily small but nonzero negative effect on the incumbent's probability of re-election. However, consistently casting unpopular votes will have a significant negative effect on an incumbent's probability of re-election. The model will not attach a specific number to this effect. Instead, we will examine the implications of its magnitude for the comparative statics of the model.

Clearly, some roll call votes may be more electorally consequential than others. The Omnibus Budget Reconciliation Act of 1993 and the Patient Protection and Affordable Care Act of 2010 stand out as obvious examples, both of which were addressed in chapter 3. However, such votes are the exceptions rather than the rule, and since the objective here is to explain the process by which legislators incrementally position themselves through the *entire* roll call voting process, such exceptions are not a fundamental challenge to the model. There are just too many other votes to explain.

4. *Party leadership candidates.* Two leadership candidates compete for the leadership position by proposing a disciplinary strategy detailing how, when, or if the party leader would pressure members on individual roll call votes. Actual leadership races can be idiosyncratic, and variable over time. As Jenkins and Stewart (2012) discuss, the evolution of parties themselves has been intertwined with the changing nature of speakership contests. One need only compare the process that brought John Boehner to the office to the one that brought Paul Ryan in as his successor to see how the practical dynamics of a leadership contest can change radically in even a short period of time. However, the purpose here is not to study the office of the Speaker nor the varied individuals who seek the office, but to use the concept of a leadership contest to study the preferences of the caucus. Thus, the model will simply treat leadership candidates the same way that Downs (1957) treated

parties. When it comes to disciplinary systems, which will be the basis of their platforms, leadership candidates are office-seekers who derive no intrinsic utility from the disciplinary strategies they propose. Hence, they will propose disciplinary systems that will win the support of the caucus, just as Downsian parties do with policy in normal elections. With respect to their policy preferences, we will assume that they have identical locations of i_{PL}, which, following from Grofman et al. (2002), puts them at some location in the mainstream of their caucus, and not at centrist locations.

We will assume that the caucus is Republican, and following that, we will assume that $i_{PL} \geq i_{MVj}$, for all districts, j. Otherwise, some districts will have median voters that are even more conservative than Republican Party leadership. While this assumption is realistic for general elections, it may be unrealistic for primaries, and chapter 7 will address how fear of "Tea Party" primaries in the post-2010 Republican caucus may have changed this dynamic. This chapter, though, will focus on general elections.

5. *The legislative agenda.* The model will make only minimal assumptions about the structure of the legislative agenda. In fact, we cannot assume conventional agenda restrictions.[3] In order to construct such rules, we must be able to summarize each legislator's complete preferences with a single functional ideal point. However, legislators in this model will sometimes behave as though their constituents' preferences are their own preferences, and they will other times behave as though their personal ideal points are their locations. Thus, we cannot construct anything like a conventional gridlock interval. However, we will make some minimal but important restrictions.

Suppose that i_{PL} represents the location of the winner of the leadership election, and i_{OPL} represents the location of the opposing party's leader. We will assume that status quo points for all bills fall within the range $[i_{OPL}, i_{PL}]$. If there were any status quo points outside that region, they would be trivially easy for the majority party leadership to change, so any votes on such status quo points should be uninteresting. Furthermore, there will be no floor votes on alternative (a) policies that are outside the range, $[i_{OPL}, i_{PL}]$. Consider the majority party leader. If $a > i_{PL}$, then the majority party leadership would not permit a floor vote on the unamended bill because even if it could pass, the majority party leadership would weaken a with amendments to bring it closer to their ideal policy. Alternatively, suppose $a < i_{OPL}$. This alternative could never reach a floor vote because it would not be reported out of committee in the first place. As a result of these restrictions, there will be no roll call votes with cut points outside the range $[i_{OPL}, i_{PL}]$. Any such vote should have a near unanimous vote within the chamber anyway.

Following from these agenda restrictions, for every vote, the majority party leader will prefer the more conservative option. Hence, *if* the majority party leader wished to apply pressure on any legislator for any vote, it would be to vote in a more conservative manner. Hence, party pressure, to the degree that it exists, will be toward conservatism rather than centrism.

While this set of agenda restrictions may seem innocuous, it is actually of vital importance, and chapter 7 will argue that the strange dynamic that developed between Republican Party leadership and its rank and file during the divided government period from 2011 through 2016 was the result of legislative agendas whose central items broke from this pattern. An agenda following Assumption 5 is an agenda under which members of the majority caucus propose legislation to move policy toward their (conservative) ideal points, away from more centrist or left-leaning status quo points. However, if the Senate or presidency is held by the opposing party and the parties are far apart, there is little reason to work on such legislation. Divided government in a *de*polarized era still makes substantive policymaking possible. President Nixon, for example, presided over a wide range of major pieces of legislation because, as a moderate, he was much closer to the similarly moderate Democratic legislative majority of the time. Under unified government, polarized though it was, President Obama enacted substantive legislation on many issues because the House and Senate majorities had policy preferences similar to his. However, when faced with a combination of divided government and polarization, major substantive legislation becomes impossible, and the incentive to engage in normal legislative policymaking disappears. With a liberal president (e.g., Obama) and a conservative House majority, there is little room for policymaking because any bill that moves policy to the right will be vetoed by the liberal president, and any bill that moves policy to the left won't make it out of the conservative Congress.

However, even when the federal government is both divided and ideologically polarized, legislation must pass in order to continue to fund the government, and more critically still, raise the debt ceiling. That means that the legislation to be addressed, though, doesn't involve a conventional status quo point following the patterns assumed here. Instead, these bills pit alternative points against "reversion" points that can fall outside the range of $[i_{OPL}, i_{PL}]$. Government shutdowns and debt ceiling breaches, for example, are often acceptable to the far-right but unacceptable to anyone else, suggesting that these outcomes are reversion points to the right of $[i_{OPL}, i_{PL}]$. That means in times that combine divided government and extreme polarization, we *can* observe bills with cut points outside that range. The model in this chapter will examine the "normal" legislative process of attempting to move policy from a status quo to an alternative, and chapter 7 will discuss what happened when the 2010 election introduced divided government into historically unheard-of levels of ideological polarization. While Assumption 5 may seem innocuous, then, it is actually critical.

The final assumption we will make about the legislative agenda is that it will include major substantive policy changes that split the parties along the central ideological divide. Recall from chapter 2 that this was the characteristic that invoked the proximity paradox.

6. *Party disciplinary tools.* The winner of the leadership contest will have unilateral authority to punish any rank-and-file caucus member, causing a legislator to lose a maximum utility of $P' > 0$, or any smaller amount. Of course, party leaders may reward loyalty as well as punish disloyalty, but this model will focus on punishments. Doing so will show that focusing on preference-preserving influence can solve the 1910 problem. In 1910, Speaker Joe Cannon faced a rebellion that resulted in his being stripped of his power in retaliation for his heavy-handed leadership style. The tale demonstrates a critical point about party leadership in Congress. Leadership positions are elected positions, so the exercise of power over the rank-and-file requires the consent of the rank-and-file. In theory, a party leader can threaten and apply harsh punishments for disloyalty, thereby achieving significant short-term policy victories. Most recently, after the 2012 election, Speaker John Boehner stripped several Republican dissidents of their committee assignments in retaliation for disloyalty on too many roll call votes, which will be addressed in chapter 7. However, applying (or even threatening) such punishments too often may alienate one's own caucus, thereby creating the same dynamic that resulted in the 1910 revolt against Speaker Cannon. In 2015, Speaker Boehner was pressured to step down. Whether that was because of his disciplinary measures or the internal divisions in the caucus is an open question, but leading through punishment is dangerous.

By focusing on punishment rather than reward, then, this model will both "stack the deck" *against* party discipline and show how party leaders can achieve dramatic effects with relatively minor punishments if they apply exclusively preference-preserving influence. Perhaps even more importantly, though, the model will derive the circumstances in which rank-and-file legislators *want* to be the recipients of such minor threats. The point is not that party discipline *necessarily* consists of punishments rather than rewards. The point is that it *can* consist of punishments without sparking rebellion, when the conditions for party government are met.

Party-imposed punishments may take a variety of forms. At the extreme end, the party leader can remove a legislator from a valuable committee assignment, as Boehner did with Representatives Justin Amash (R-MI), Tim Huelskamp (R-KS), Walter Jones (R-NC), and Dave Schweikert (R-AZ) shortly after the 2012 elections. Given reappointment norms in Congress, these are severe punishments, reserved for dramatic occasions. More mild punishments might consist of allowing a minor bill sponsored by the legislator to languish without action, or reducing the amount of pork flowing to the legislator's district. At the lowest level of severity, a punishment might not even be associated with a specific action by party leaders. Party leaders might simply tally legislators' refusals to defer to leaders, and treat each refusal as a "demerit." Earning a single demerit may have no consequence at all, but party leaders may simply increase the probability of a tangible punishment as a legislator accrues more demerits. After accruing a sufficient number of demerits, requests for committee transfers may be denied, or other such real consequences may follow. Thus, the cost of refusing to defer to leadership may be as small as giving up the opportunity to do so later without direct consequences. A punishment may even be as simple and modest as a degree of ostracism. What is important is that there is variation in the severity of threats that parties can make, from the harsh to the innocuous, as discussed by Pearson (2015).

Keep in mind, though, that the structure of the model would work equally well if party leaders merely have the potential to dole out modest rewards rather than punishments. Just as punishments can range in magnitude from severe to innocuous, so can rewards. A reward at the extreme positive end could be a plum committee chairmanship, but, obviously, leaders have limited ability to dole these out, just as they can only strip so many legislators of their committee assignments before provoking another 1910-style revolt. Similarly, a reward can be as modest as praise and inclusion, which is the analog to punishment by ostracism. Either way, the smaller the incentive, the more party leaders can use the incentive, punishment, or reward.

7. *Information*. Legislators know each others' policy preferences and district median voters' locations.

Analytic Approach

Equilibria to the model will consist of a pair of leadership platforms, each legislator's vote in that contest, the sequence of roll call votes, and the subsequent round of elections. Deriving a *specific* equilibrium, though, would require the set of initial status quo points for each policy. Since they cannot be derived from first principles, the chapter will focus on equilibrium conditions. Specifically, the chapter will derive the conditions under which a laissez-faire leadership platform is defeated by a platform of applying preference-preserving influence. When these conditions are met, the caucus will prefer a leader who applies preference-preserving influence to a laissez-faire leader, so the absence of preference-preserving influence will be precluded in equilibrium. The chapter will then trace out the policy and electoral effects of a caucus selecting a leader who applies preference-preserving influence, and then discuss why the conditions under which this occurs preclude preference-deviating influence.

Two Leadership Platforms

While there are many variations of disciplinary systems a leadership candidate might propose, the goal is merely to show that when important conditions are met, the absence of discipline cannot occur in equilibrium because the caucus wants preference-preserving influence. There are four basic categories of disciplinary systems: a laissez-faire system with no disciplinary restrictions, a preference-preserving system, a preference-deviating system, and a mix of the latter two. The analysis will compare a completely laissez-faire leadership style to the simplest form of preference-preserving influence and derive the conditions under which the caucus prefers the latter to the former. Then we will examine the conditions under which a leader might successfully apply preference-deviating influence, and show both that such disciplinary systems are unsustainable on a large scale and that when the "conditions for party government" are met—meaning that a preference-preserving system defeats a laissez-faire system—there are limited opportunities for preference-deviating influence anyway. When the conditions for party government are met, then, office-seeking leadership candidates must offer systems of preference-preserving influence.

Laissez-faire leadership: A laissez-faire platform is one that promises to allow caucus members to vote however they choose with no rewards or punishments.

Preference-preserving leadership: Given Assumption 5, for all roll call votes, a party leader will prefer the conservative option to the liberal option. The leader's action will depend on which circumstance she faces on any given roll call vote, (a), (b), or (c).

(a) If every caucus member votes consistently with their constituents' preferences, the conservative option will win.
(b) If every caucus member votes consistently with their constituents' preferences, the liberal option will win, but if every caucus member votes sincerely, the conservative option wins.
(c) If every caucus member votes consistently with their constituents' preferences, the liberal option wins, and if everyone votes sincerely, the liberal option still wins.

If a preference-preserving leader finds herself in circumstance (a), she will apply no discipline. By logic analogous to Groseclose and Milyo (2010), the conservative option will win in the absence of pressure, so she has no need to apply any pressure. If she finds herself in circumstance (c), she will permit her caucus to vote for the option their constituents prefer because enforcing sincere voting would accomplish nothing except potentially to alienate the rank-and-file and their constituents. The caucus simply isn't sufficiently conservative for the party leader to be able to impose her preferences.

However, if she finds herself in circumstance (b), she will do as follows. Let prob(V | sincere or insincere) denote the conditional probability that a legislator will be re-elected given either a sincere or an insincere roll call vote. If any caucus member, *j*, sincerely prefers the conservative option, a preference-preserving leader threatens to impose a punishment, P > 0, on that legislator if she votes insincerely, where P meets the following condition:

$$P > V\left|\left[\text{prob}\,(V\,|\,\text{insincere vote}) - \text{prob}\,(V\,|\,\text{sincere vote})\right]\right| - \left|U_j(\text{sq}) - U_j(\text{a})\right| \quad (4.1)$$

The condition in Inequality 4.1 simply requires that the punishment is large enough to outweigh the electoral cost of voting sincerely on the bill in question. By Assumption 3, |[prob(V | insincere vote) – prob(V | sincere vote)]| is quite low for any single arbitrary vote. Thus, by Assumption 3, even if V is large, the right-hand side of Inequality 4.1 will be small. Hence, the punishment required to induce a sincere vote will be quite small with the exception of the occasional high-profile vote. In fact, if the legislator expects to be pivotal, such punishment may be unnecessary to induce a sincere vote. If the legislator expects to be pivotal, the punishment required would be as follows:

$$P > V\left|\left[\text{prob}\,(V\,|\,\text{insincere vote}) - \text{prob}\,(V\,|\,\text{sincere vote})\right]\right| - \left|U_j(sq) - U_j(a)\right| \quad (4.2)$$

For reasons discussed above, $V\left|\left[\text{prob}\,(V\,|\,\text{insincere vote}) - \text{prob}\,(V\,|\,\text{sincere vote})\right]\right|$ will be quite small. Furthermore, as chapter 3 demonstrated, $\left|U_j(sq) - U_j(a)\right|$ may be unboundedly large. Therefore, the right side of Inequality 4.2 will generally be negative, so if a legislator expects to be pivotal, a preference-preserving leader might not need to threaten *any* punishment to induce a sincere vote. The point is that the prospective leader only threatens punishments large enough to induce sincere votes, and even very modest threats can induce sincere votes. This will become critical when we compare preference-preserving influence to preference-deviating influence later in this chapter. The preference-preserving leader only attempts to induce sincere votes, and only when doing so solves the collective action problem of sincere voting.

Legislative and Electoral Consequences of Voting for Either Leadership Platform

Given the disciplinary platforms detailed above, we can trace out the effects of electing a leader with that platform on the subsequent stages of the game. Suppose that one leadership candidate runs on a platform of laissez-faire leadership and the other runs on a platform of preference-preserving influence, as detailed above.

Under laissez-faire leadership, roll call voting would reduce to Groseclose and Milyo (2010). All members defer to their constituents unless doing so creates a margin of precisely one vote, in which case the result is indeterminate. Hence, a vote for laissez-faire leadership is an acceptance of the collective action problem of sincere voting. What would that mean for the positions that legislators incrementally adopt, and what would that imply about election results?

For any individual roll call vote, the electorally optimal vote is always the vote that is consistent with the preferences of the district median voter. If a legislator always votes in accordance with the preferences of the district median voter, then that is equivalent to incrementally adopting the location of the median voter. Under laissez-faire leadership, then, legislators representing districts with extreme preferences will incrementally adopt extreme positions, and legislators representing centrist districts will incrementally adopt centrist positions. Empirically, then, we would observe a degree of ideological cohesiveness in voting scores within the caucus dependent entirely on the degree of *electoral* homogeneity within the caucus, with negligible deviations between any one legislator's voting record and the preferences of her constituents. Electorally, then, caucus members should pay essentially no cost for their voting records because their voting records would be tailored to the preferences of their constituents.

How does that compare to the positioning process and subsequent elections under preference-preserving influence? Under preference-preserving influence, rank-and-file legislators are free to defer to their constituents only if this produces the same legislative outcome as mandatory sincere voting. If not, the leader threatens to punish any caucus member who personally prefers the conservative option, but is tempted to vote insincerely for electoral reasons. The threat is sufficient to achieve compliance, so the punishment will not be carried out. Thus, for any vote on which sincere voting is a collective action problem, while a laissez-faire leader permits unlimited free-riding, the preference-preserving leader permits none.

What does that mean for the incremental positioning process under preference-preserving leadership? The answer comes from Assumptions 4 and 5.

By Assumption 4, the leader has a more extreme location than any given district median voter. Thus, consider her relationship with an arbitrary rank-and-file legislator, located at MC, whose district median voter is located at MV. The leader's platform is such that she will never threaten punishment except when casting an electorally optimal roll call vote yields a roll call vote inconsistent with leadership's preferences. Moreover, she will never attempt to induce an insincere roll call vote. Thus, for any roll call vote under which an arbitrary member of Congress would behave differently under the two platforms, the following conditions must hold: $MV < (l + c) / 2 < PL$ and $(l + c) / 2 < MC$.[4] When party leadership applies pressure under these conditions, the member of Congress will respond by voting for the conservative option, *c*. Since we *estimate* legislators' locations based on their roll call voting patterns, the *occurrence* of these roll call votes will result in an estimate of the legislator's location that is more ideologically extreme than the location we would have observed under laissez-faire leadership, which would have resulted in the member of Congress voting for *l*, as a moderate would have done.

More broadly, we would observe less responsiveness to district preferences when parties operate as preference-preserving institutions precisely because they function to cancel out constituent influence. Thus the degree of electoral heterogeneity or homogeneity within the caucus (similarity between legislators' districts) would be less predictive of homogeneity in roll call voting patterns because district preferences would be less predictive of representatives' roll call voting patterns. Instead, the degree of cohesiveness within the caucus will depend more on the distribution of *personal* ideal points within the caucus because the preference-preserving leader would reduce the frequency with which caucus members cast insincere roll call votes. Our *observation*, though, would be that legislators incrementally adopt more extreme positions in the policy space, deviating from the preferences of their constituents in the rightward direction, given the assumption of a right-leaning party (and obviously, in a leftward direction were we to examine a left-leaning party).

What would that mean for legislative elections under preference-preserving influence? At the most basic level, it would mean that caucus members would *not* build voting records that maximize their vote shares. However, this does *not* mean that they lose. Even if the resulting bundle of roll call votes reflects a location further from the district median voter than the bundle that the challenger would have built, a proximity paradox could still hand the election to the incumbent. That simply requires a legislative agenda with items that move policy a significant distance while cleanly dividing the parties along the central ideological break, as we saw in chapter 2, and we include such agenda items by Assumption 5 since such agenda items are reasonable. With the inclusion of such agenda items, extremists in *both* parties can defeat more centrist challengers, or at least challengers expected to be more centrist given party reputations. Incumbents simply win by smaller margins than if they had incrementally adopted more centrist locations.

So laissez-faire leadership will result in legislators failing to achieve policy goals, adopting centrist locations when representing centrist districts, and generally maximizing their vote shares. When legislators establish parties as *preference-preserving institutions,* though, they will achieve policy goals, adopt more extreme positions, and fail to maximize their vote shares, but still frequently be re-elected by narrower margins. With that information, we can now derive the conditions under which the legislative caucus will choose preference-preserving leadership over laissez-faire leadership, thereby deriving the conditions for party government.

Caucus Preferences over Leadership Platforms

The conditions for party government when parties function as preference-preserving institutions are the conditions under which the caucus will prefer the preference-preserving leader to the laissez-faire leader. Under what circumstances will an arbitrary member, *j*, prefer the former to the latter even though she pays electoral costs?

Member *j*'s preference ranking between the two hypothetical leadership platforms depends on the total electoral cost of voting to avoid the activist leader's punishments, and the policy consequences. Let M represent the set of votes on which Member *j* would pay an electoral cost to avoid punishment. Let S be the set of votes on which the party leader's discipline would cause the conservative option to win when it would otherwise lose. Let N represent the subset of S on which Member *j* sincerely prefers the conservative option, and let Q represent the subset of S on which Member *j* sincerely prefers the liberal option. Note that we do not need to include the number of times Member *j* will be punished. The party leader's strategy is to never threaten punishment unless the threat will be successful, so the punishment will never be carried out.

Member *j*'s preference between the hypothetical leadership platforms depends on a simple cost-benefit analysis. Each time the Member of Congress casts an electorally suboptimal roll call vote to avoid a punishment from the activist leader, she pays an electoral cost, incrementally reducing her probability of re-election. The total cost to her, from that electoral penalty in utility terms, then, is given by the following series.:

$$V\sum_{m=1}^{M}|\, prob(V\,|\,sq_m) - prob(V\,|\,a_m)\,| > 0 \qquad (4.3)$$

The utility that Member j would gain by receiving preferred policy outcomes under the activist leader's authority is given by the following:

$$\sum_{n=1}^{N} |U_j(sq_n) - U_j(a_n)| > 0 \tag{4.4}$$

Finally, the utility that Member j would lose by receiving her least preferred policy outcomes under activist leadership is given by the following:

$$\sum_{q=1}^{Q} |U_j(sq_q) - U_j(a_q)| > 0 \tag{4.5}$$

Thus, Member j will prefer a leader who applies preference-preserving influence to the leader who applies no influence if the following inequality holds:

$$V \sum_{m=1}^{M} |\, prob\,(V \,|\, sq_m) - prob\,(V \,|\, a_m)\,| < \sum_{n=1}^{N} |U_j(sq_n) - U_j(a_n)| - \sum_{q=1}^{Q} |U(sq_q) - U_j(a_q)| \tag{4.6}$$

Inequality 4.6 tells us what must be true for a given member of Congress to want a party leader who applies preference-preserving influence to solve the collective action problem of sincere voting. We can generalize from Inequality 4.6 to discuss what must be true of a caucus in order for a majority to prefer activist leadership to laissez-faire leadership. For any fixed utility of re-election, we can impute values to the left-hand side of Inequality 4.6 based on the empirical estimates of Canes-Wrone et al. (2002), but we cannot evaluate the magnitude of any element in either series on the right-hand side because, as chapter 3 demonstrated, their values are indeterminate. However, we can evaluate Inequality 4.6 by examining the *number* of elements in each series since each element is positive. Given that, the conditions for the caucus to elect the party leader who applies preference-preserving influence are as follows:

(a) *Ideological homogeneity*. If the caucus is ideologically homogeneous, meaning that members' personal ideal points are clustered near leadership's location, then for any one member, the second series on the right-hand side of Inequality 4.6 will have very few elements because there will be very few votes on which they disagree with leadership. Thus, the second series on the right-hand side of Inequality 4.6 will be small, making Inequality 4.6 more likely to be true. This is rather obvious, and consistent both with traditional CPG and the standard criticism of that theory.

(b) *Electoral heterogeneity.* Electoral heterogeneity means that caucus members represent districts with a wide range of median voters. The combination of ideological homogeneity and electoral heterogeneity is what creates the collective action problem of sincere voting in the first place.[5] The elements in the first series on the right-hand side of Inequality 4.6 are votes on which the member and most of her colleagues prefer the conservative option, but enough members' constituents prefer the liberal option to cause the conservative option to fail in the absence of party influence because of electoral deference. Assuming ideological *homo*geneity, the more electorally *hetero*geneous[6] the caucus is, the more elements there will be in that series, making Inequality 4.6 more likely to be true for any given member.[7] After all, if every caucus member represented a district with noncentrist preferences that match the caucus members' actual policy preferences, there would be no collective action problem. It is only when there is a contingent within the caucus who represent districts that *don't* share those preferences that a collective action problem exists.

(c) *Strong policy preferences.* The right-hand side of Inequality 4.6 represents the policy gains a member receives by having the party leader solve the collective action problem of sincere voting. Assuming conditions (a) and (b) are met, the right-hand side of the inequality will be large. However, in order for Inequality 4.6 to be true, these gains must be larger than the total electoral cost of the forced sincere votes. While one roll call vote is unlikely to cost a legislator her seat, a pattern of unpopular votes might, by Assumption 3. As Canes-Wrone, Brady, and Cogan demonstrate, a consistently "out-of-step" member of Congress can easily lose five percentage points from her vote share at the end of the term, and while that isn't enough to cost most incumbents their seats, it may be enough to cost members from close districts their seats, and they are precisely the members who are most likely to be subject to preference-preserving influence from leadership, because any legislator whose constituents take the same position as leadership would be compelled by electoral pressure to do what leadership wants anyway, with or without party influence. It is the members from close districts who need preference-preserving influence in the first place. Moreover, as shall be addressed in chapter 5, there are more members from marginal districts than conventional wisdom suggests.

A regular pattern of extreme roll call voting is electorally suboptimal. Hence a legislator will only support a leader who applies preference-preserving influence if she is willing to sacrifice some electoral security in order to achieve policy victories. The critical point is that members of Congress are rarely pivotal on roll call votes. Hence, in the absence of parties, they are rarely forced to confront the choice between policy

> victories and electoral security. If rational, then, they just choose electoral security. However, the choice between preference-preserving influence and laissez-faire leadership forces the caucus to confront that choice directly, and members will only choose activist leadership if they would be willing to sacrifice some electoral security for policy victories. Hence the final condition for party government in the preference-preservation model is that the caucus weighs policy outcomes sufficiently heavily to compensate for the electoral costs of being forced to vote sincerely, thereby incrementally adopting ideologically extreme, electorally suboptimal positions in the policy space.
>
> We must remember a critical result from chapter 3, though. The potential cost of a policy loss, and hence the potential benefit of a policy victory, can be unboundedly large, even when a legislator's ideal point is relatively close to the cut point for the roll call vote. Thus, the policy motivation can be arbitrarily strong, hence the incentive to empower party leaders to enforce sincere voting.

Note also that since Inequality 4.6 abstracts away from the policy space, the logic is equally applicable to multidimensional models and to unidimensional models. The basic decision rule for any legislator's vote is to compare the policy costs of party discipline, the policy benefits of party discipline, and the electoral costs of being pressured to cast electorally damaging votes. These remain the same regardless of the number of policy dimensions. Appendix A elaborates.

These three conditions—ideological homogeneity, electoral heterogeneity and strong policy motivations—are the conditions under which the *absence* of party discipline *cannot* occur in equilibrium because when these conditions are met, the caucus will prefer a leader who applies preference-preserving influence to a leader who applies no influence. These conditions for party government, while somewhat different from their counterparts in existing variations of CPG, should not be surprising. The ideological homogeneity condition, in particular, creates a direct connection between this model and conventional CPG. Furthermore, the relative weights assigned to policy and electoral considerations should be obvious. If party leaders solve a collective action problem to provide policy at an electoral cost, then the trade-off will only be attractive to members of Congress if they care strongly about policy.

The most interesting condition is the electoral heterogeneity condition. Members of Congress will prefer strong party leadership if they are ideologically homogeneous but electorally diverse because that is precisely what creates the collective action problem in the first place. If they are ideologically heterogeneous, they will not perceive the policy effects of party leadership to be a collective good because they will not perceive the effects to be *good*. If they are electorally homogeneous, they will have no need for external pressure to ensure unified voting, so there will be no need for party leadership. It is the combination of ideological homogeneity and electoral heterogeneity that creates incentives for members of Congress to empower party leaders to enforce discipline.

Thus, a central prediction of this model is that an ideologically homogeneous caucus *without* a strong party will only vote cohesively if its members represent similar districts, which they never actually do. If they don't have similar districts (and empirically, they don't), then an ideologically homogeneous caucus can *only* build cohesion by empowering party leaders to solve the collective action problem of sincere voting. This eliminates any semblance of tautology in CPG since it implies that ideological homogeneity is *not* a sufficient condition for cohesion on roll call voting. An ideologically homogenous caucus without a strong party will *not* vote the same way as an ideologically homogenous caucus *with* a strong party.

This model does, however, conflict to some degree with the Cartel model. Cox and McCubbins (1993, 2005) argue that parties provide their members with collective electoral benefits by creating a "brand name" through party unity. Relatedly, Sniderman and Stiglitz (2012) argue that politicians have incentives for party consistency in order to secure the "reputational premium" of that party. In contrast, this chapter argues that parties provide their members with collective policy benefits while imposing electoral *costs*. Moreover, there is no need in this model for Lebo et al.'s (2007) party-wide electoral benefit of legislative success to compensate for the individual electoral costs of deferring to party leadership. Without those benefits, the extreme positioning that results from accepting preference-preserving influence has both direct and indirect electoral consequences. It encourages challengers (Carson 2005), and reduces incumbents' expected vote shares directly (Canes-Wrone et al. 2002). That latter effect happens incrementally, but those incremental effects add up. The effect is akin to the problem of eating potato chips. One potato chip will have a negligible effect on one's weight, but eating them in large numbers will have consequences. Parties in this model are similar to the proverbial devils on our shoulders telling us that one more potato chip won't do any harm. One electorally damaging vote usually won't cost the incumbent her seat, but sequentially casting electorally unpopular votes based on the marginal electoral costs, one vote at a time, can have ramifications because the marginal effects add up. That is the point of the balance between electoral and policy considerations for each member of Congress. It is rational for legislators to empower leaders to enforce sincere voting if and only if the *aggregate* policy benefit of giving leaders the necessary tools outweighs the *aggregate* electoral cost of all the sincere, electorally suboptimal votes that they are forced to cast.

Equilibrium Conditions and the Explanation for Extreme Positioning

The conditions under which a majority of the caucus would choose a preference-preserving disciplinary system over a laissez-faire system are the conditions under which the former defeats the latter in a leadership contest, in which case party influence *must* exist in equilibrium. Moreover, if the caucus is ideologically homogeneous, then by necessity, there will be very few occasions on which a party leader could even potentially apply preference-deviating influence because ideological homogeneity implies that the caucus shares the leader's noncentrist policy preferences. We will examine the possibility of preference-deviating influence later in this chapter, but for now, the critical observation is that when the caucus is ideologically homogeneous and electorally diverse, sincere voting will be a collective action problem, and when caucus members ascribe sufficient utility to policy outcomes relative to electoral security, they will *want* a leader to solve that collective action problem. Thus, they will elect a leader who promises to apply preference-preserving influence.

Based on Assumption 4, each time a leader applies preference-preserving influence in order to solve the collective action problem of sincere voting, the result is that each pressured legislator casts a roll call vote that contributes to a voting score, that is, an observed policy location, that is more ideologically extreme than the location that she would have adopted in the absence of party influence. Thus, we begin to see the outlines of an answer to the final critical puzzles motivating the book—why do legislators adopt such ideologically extreme, electorally suboptimal positions, and what role to parties play? The answer is that legislators empower party leaders to prod them into incrementally adopting extreme, but sincere positions. Doing so allows the caucus to achieve policy gains that they could not otherwise achieve because of the collective action problem of sincere voting.

However, several questions remain about equilibrium conditions. Two, in particular, stand out. How much electoral damage will be done by empowering party leaders in this manner, and can preference-*deviating* influence occur in equilibrium? The remainder of this chapter will explore these questions.

Mitigating the Electoral Costs of Preference-Preserving Influence with the Agenda

The problem with empowering party leaders to impose electorally costly discipline is that such discipline could, in principle, cost the majority party its majority status by costing its members their re-elections. If the caucus members are individually policy-motivated, then by definition they are willing to absorb some electoral risk for policy gain, but the analysis in chapter 2 suggests that the risk can be quite modest because while party discipline would force legislators to adopt electorally suboptimal positions, a normal legislative agenda will mitigate those electoral effects. To be sure, the application of preference-preserving influence precludes the possibility of legislators adopting electorally *optimal* positions. These agenda effects will only minimize the nonzero electoral costs.

Consider an arbitrary member with extreme personal preferences and a centrist district median voter. For demonstration purposes, suppose that the district median voter is located at .1, and that the legislator's personal ideal point is .5. In principle, a legislative agenda does not have to include any votes with cut points in the range [.1, .5], in which case the legislator would pay no electoral cost for sincere voting. However, if the conditions for party government are met, then she is willing to pay at least some electoral cost in exchange for policy victories, so if it would guarantee policy victory, she would be willing to cast at least some sincere roll call votes with cut points in that range. Moreover, if there were no roll call votes with cut points in that range, the legislator's location would be empirically indistinguishable from .1 anyway. Alternatively, an agenda with very few roll call votes with cut points in the [.1, .5] range would make it possible to distinguish empirically between locations of .1 and locations of .5, but minimize the electoral damage of those differences, as discussed in chapter 2. However, again, a policy-motivated legislator is willing to pay electoral costs for policy victories. Thus, we have no a priori reason to expect the agenda to be devoid of votes with cut points in that region. There is, however, one more agenda structure that can mitigate the electoral costs imposed by party discipline while still providing significant policy victories on votes with cut points in the [.1, .5] range. If the agenda is *also* stacked with votes whose cut points are at least slightly to the left of .1, which no leader has any incentive to *block* because they wouldn't divide the majority caucus, then that is sufficient to invoke the proximity paradox.

Each roll call vote with a cut point to the right of the challenger's location and to the left of .1 *adds* relative utility to the bundle of roll call votes the incumbent offers to a majority of the electorate, and *subtracts* relative utility from the bundle of roll call votes the *challenger* offers to a majority of the electorate. So unless the challenger adopts the Condorcet-winning location of .1, which this chapter prohibits by assuming that party reputation constrains challengers, the majority caucus can increase members' vote shares merely by stacking the legislative agenda with *additional* items on which sincere roll call votes are electorally *popular.* So the majority caucus can counteract at least some of the electoral damage done by incumbents' sincere but *un*popular roll call votes, thereby creating the proximity paradox derived in chapter 2.

We should expect such agenda items anyway. Recall from chapter 2 that the structures that create the proximity paradox beneficial to incumbents of both parties are the votes that separate the parties cleanly along the central ideological divide. Votes with cut points slightly to the left of .1 do precisely that.

Combining agenda items with cut points in the [.1, .5] range with party-enforced sincere voting, our hypothetical legislator will not cast a series of roll call votes that *maximizes* her vote share. Thus, we should still observe a negative effect of incumbent extremism on vote shares in multivariate analysis, but that effect can be arbitrarily small as long as it is counteracted by a set of party-line votes that benefit Democratic extremists in even slightly Democratic-leaning districts, and Republican extremists in even slightly Republican-leaning districts. We should observe such votes anyway because they are the votes that split the parties along the central ideological divide. The incumbents who *suffer* from these votes are the mismatched incumbents, such as any Democrat representing an even *slightly* Republican-leaning district (see, for example, the Democrats who won seats in the 2006 and 2008 wave elections, who paid electoral penalties for their votes in favor of the Patient Protection and Affordable Care Act in the 2010 midterm election). However, a liberal extremist will still defeat a conservative-leaning moderate in an at least slightly liberal district, and a conservative extremist will still defeat a liberal-leaning moderate in an at least slightly conservative district.

The electoral paradoxes derived in chapter 2, then, come to the rescue of the legislative caucus. A caucus that is ideologically homogeneous, electorally diverse, and policy motivated will empower party leaders to solve the collective action problem of sincere voting, thereby forcing rank-and-file members to incrementally adopt sincere, extreme, and electorally suboptimal locations in the policy space, but the structure of the legislative agenda will reduce the damage of extreme positioning because normal legislative agendas, as discussed in chapter 2, can permit extremists to defeat centrists. Thus, when the conditions for party government are met, we will observe polarization in voting scores and weak negative effects of ideological extremism on vote shares in legislative elections, but many extremists will survive their re-election campaigns anyway because of the weakness of that effect.

Variations of Preference-Preserving Influence That Minimize the Electoral Costs

The analysis so far has been based on a hypothetical leadership contest between a laissez-faire leader and one who pledges to apply preference-preserving influence in one specific manner, but the variations of potential disciplinary systems are limited only by the size and structure of the legislative agenda. However, two alternative platforms merit discussion as ways that also mitigate the costs of party discipline. First, a party leader might apply preference-preserving influence on a more limited subset of votes than the strategy detailed above. After all, some votes are more electorally consequential than others, and some have less salient policy effects than others. On such votes, members might not want to pay the electoral cost of a policy victory, so a strategic party leader might not apply any pressure. The other obvious variation would be to apply preference-preserving influence, but only to reach a bare majority. Such a strategy permits some free-riding, but not enough to affect the policy outcome, and would be more compatible with Snyder and Groseclose's (2000) argument that parties' influence may be stronger on close votes because that is when they have the most incentive to apply pressure. While there are nearly limitless variations on the selective application of preference-preserving influence, the structure of the optimal variation is unimportant. The basic structure of any one member's preference between preference-preserving influence and no party influence will still be given by Inequality 4.6. All that would change for variations on the preference-preservation platform would be the specific elements in sets M, N, and Q. Hence, the same conditions would create incentives for the caucus to choose preference-preserving influence: ideological homogeneity, electoral heterogeneity, and strong policy preferences. When those conditions are met, the caucus will want party leaders to apply preference-preserving influence. The result will be the incremental adoption of extreme positions with modest electoral costs that are mitigated by the structure of the legislative agenda.

Preference-Deviating Influence

Finally, then, we come to the crux of the party influence question. The analysis so far has been based on a comparison between a platform of preference-preserving influence and a laissez-faire platform. Can preference-deviating influence occur in equilibrium and, if so, how often? Clearly, there are circumstances in which a strong majority of a caucus could prefer for a party leader to apply at least some preference-deviating influence. Suppose that a caucus consists of 215 legislators with personal policy preferences equivalent to a DW-NOMINATE score of .6, and 15 legislators with preferences of .1. Suppose, further, that the minority caucus consists of 205 legislators, all of whom have preferences equivalent to −.6. In order to win any roll call vote with a cut point in the range of [.1, .6], the majority party leadership would need to apply preference-deviating influence to at least three members of the moderate contingent of the majority caucus. The 215 conservatives in the caucus would be happy to have the leadership do so since they receive policy benefits. Thus, any time such a vote comes up, the majority party leadership can choose three moderates to pressure into an insincere vote. Those moderates cannot form a large enough coalition within the party to defeat the leader in a leadership contest, so as long as the party leader doesn't alienate the moderates so much that more than 12 of them switch parties or push for a new leader under that threat, the party leader will maintain her position and keep 215 of her 230 caucus members very happy through the occasional and selective application of preference-deviating influence. In fact, those 215 conservatives will be happier than if the leader refused to apply any preference-deviating influence. The only constraint is the willingness of the 15 moderates to retaliate.

Clearly, then, the model presented here does not preclude the application preference-deviating influence in every circumstance. However, several points must be made about that type of party influence. First, members who are subjected to preference-preserving influence can actually *prefer* to be collectively pressured if they ascribe enough weight to policy. However, members who are subjected to preference-deviating influence *cannot* gain from being pressured. Those members cannot benefit policy-wise because, by definition, they don't like the policy result. Furthermore, they cannot benefit electorally because the motivating principle of this model is that in the absence of party influence, members of Congress should vote in the most electorally beneficial way anyway. Since nobody can benefit from being subjected to preference-deviating influence, its application necessarily runs counter to the will of *some* of the caucus, which limits the circumstances in which it can be applied as long as leaders must maintain support within the caucus. Finally, recall that if a caucus meets the conditions for party government, then by definition there should be relatively few instances in which caucus members would have preferences that conflict with leaders' preferences anyway.

When evaluating the feasibility of preference-deviating influence, then, we must address two questions—what is necessary for it to be successful, and how common are the required circumstances? Then we can consider the practicality of leadership candidates incorporating preference-deviating influence into their platforms. As we shall see, the requirements for preference-deviating influence to be successful are either harsh punishments, capricious threats, or unusual conditions. That fundamentally limits the feasibility of widespread preference-deviating influence. On the other hand, preference-preserving influence has reasonable circumstantial requirements and only requires minimal punishment, making it far more appealing to would-be leaders crafting their platforms.

In order to demonstrate this, consider Table 4.1, showing every possible interaction between legislators' preferences, constituents' preferences, and leadership preferences, describing what type of influence a leader might want to apply in that circumstance.

If a member's sincere preferences are consistent with both the party leader's preferences and her constituents' preferences (Case 1), the member of Congress will vote consistently with the leader's preferences with or without pressure. That leaves three circumstances in which the party leader might want to exert influence. If the legislator's preferences are inconsistent with the party leader's preferences, then influence would be preference-deviating influence (Cases 2 and 3). However, if the legislator's preferences are consistent with the party leader's preferences and inconsistent with her constituents' preferences, influence would be preference-preserving influence (Case 4). Consider each in turn.

Case 1

If both a legislator's sincere policy preferences and her constituents' preferences are consistent with the party leader's preferences, then the party leader does not need to apply any pressure in order to induce party loyalty.

Case 2

Suppose an arbitrary legislator faces a vote on which she agrees with her district median voter, but not her party leader. By Assumption 4, we must place the party leader to the right of the median voter. With that restriction, there are two possible arrays that can generate Case 2, given by Figures 4.1 and 4.2.

In Figures 4.1 and 4.2, the legislator in question prefers the liberal option to the conservative option. In the absence of party influence, she will vote for the liberal option since that is her sincere preference and the most electorally beneficial vote. Since that is inconsistent with the party leader's preference, the party leader might be tempted to use party discipline to induce an insincere vote. How strong must her disciplinary measures be to do so?

The magnitude of punishment necessary to induce an insincere vote depends on the legislator's belief about whether or not she is pivotal. If the legislator believes that her vote will be pivotal, then in order for the party

Table 4.1 **Combinations of preferences and influence type**

		District median voter's prefs.	
		Consistent w/ MC prefs.	*Inconsistent w/ MC prefs.*
Leader's prefs.	Consistent w/ MC prefs.	(1) No influence	(4) Preference-preserving
	Inconsistent w/ MC prefs.	(2) Preference-deviating	(3) Preference-deviating

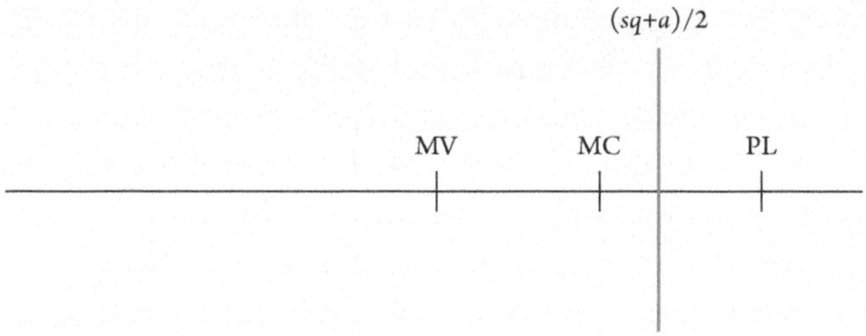

Figure 4.1 Case 2: Preference-deviating influence, variation 1

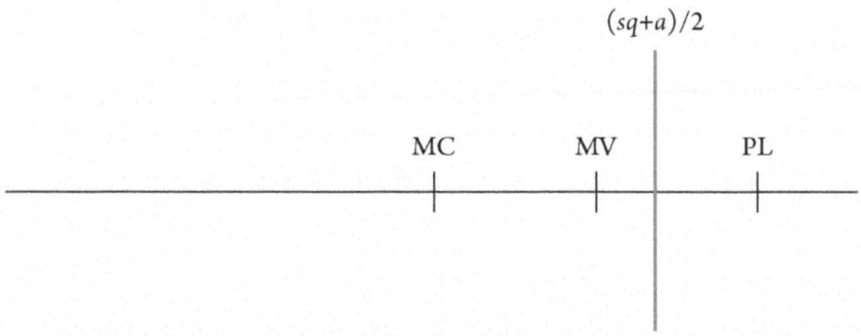

Figure 4.2 Case 2: Preference-deviating influence, variation 2

leader to induce an insincere vote, she must offer incentives that counteract both the electoral damage of voting for the conservative policy, and the utility differential between the liberal and conservative policies. Hence, the punishment must meet the following condition:

$$P > [U(l) - U(c)] + V\left[\text{prob}(V|\text{liberal vote}) - \text{prob}\left(V\middle|\text{ conservative vote}\right)\right] \quad (4.7)$$

A very large punishment would be required to satisfy Inequality 4.7. First, as chapter 3 showed, $|U(sq) - U(a)|$ may be unboundedly large, even when the cut point is arbitrarily close to the legislator's ideal point, so simply threatening a punishment large enough to counteract the policy cost of an insincere vote is challenging enough. However, the leader's punishment must be even greater than that because it also has to counteract the utility the member of Congress loses by incrementally reducing her probability of victory in the next election through the construction of an electorally suboptimal bundle of roll call votes associated with a more extreme location in the policy space. By Assumption 3, that additional cost is small. However, the fact that the party leader must counteract both policy and electoral incentives to induce an insincere vote means that she must use a rather severe threat to induce an insincere vote if the legislator expects to be pivotal. A legislator subjected to such pressure will likely be rather unhappy about it since she is being pressured to cast an electorally damaging vote that results directly in a policy outcome that she sincerely personally dislikes. Now let us reconsider the case of Representative Marjorie Margolies-Mezvinsky (D-PA) and the 1993 Omnibus Budget Reconciliation Act. If party pressure induced an insincere vote, it must have been quite a threat.

What does that mean from the perspective of a strategic party leader? First, consider the fact that a House bill requires 218 votes to pass (with full participation). So if a party leader is scrounging for an additional vote to get above 217 (recall that we are considering the case of a pivotal legislator), she might consider attempting to pressure someone who would require a less severe threat, or who would be less aggrieved by the threat. It is possible that the party leader cannot find a more persuadable legislator, in which case she must weigh the benefits of the policy victory against the cost of alienating the caucus member in question. As the example at the beginning of this section demonstrated, the party leader may actually increase her overall support within the caucus by applying preference-deviating influence in this case, but the important point for the moment is that the decision is not without cost.

Alternatively, now, suppose the member of Congress believes that her vote will not be pivotal. Inducing an insincere roll call vote only requires a punishment large enough to outweigh the electoral damage of a single unpopular roll call vote, and by Assumption 3, that isn't very large. However, if the legislator's vote will not be pivotal, then there is no reason for the party leader to apply such pressure. Nobody would receive any policy benefit because the legislator's vote is not pivotal, just as the member of Congress will pay no policy cost because the vote is not pivotal. However, the party leader will have forced the legislator to cast an electorally damaging roll call vote for no apparent reason, which may create enmity through capriciousness, as well as putting a seat at risk. The analysis earlier demonstrated that legislators may tolerate being asked to cast electorally damaging sincere votes when those votes won't be pivotal if they are simply being prevented from free-riding. However, if they disagree with the policy for which they are being asked to vote, there is no reason for them to tolerate such pressure. Hence, a strategic party leader has no incentive to apply preference-deviating influence to a nonpivotal legislator.

Regardless of whether or not a legislator expects to be pivotal, then, applying preference-deviating influence in Case 2 creates potential problems for the party leader. Either the pressured legislator will be upset because of the harsh disciplinary measures threatened and the combination of policy and electoral damage caused, or because of the capricious nature of the threat. In either case, the party leader probably cannot count on the pressured legislator supporting her in a leadership contest.

Most importantly, though, if such circumstances occur with frequency, that implies that the caucus is not ideologically homogeneous because roll call votes compatible with Case 2 place the member of Congress on the opposite side of the cut point from the party leader. If such circumstances are common, then the conditions for party government are not met anyway, so the caucus will not tolerate strong leadership under any model.

Case 3

Next consider the case in which the leader's preferences and the legislator's constituents' preferences are consistent with each other, but inconsistent with the legislator's preferences. The party leader may consider applying preference-deviating influence, and it would seem to be the easiest circumstance to do so. When party pressure and electoral pressure work together, party pressure does not need to be as strong to influence roll call votes because it does not work alone. In fact, if the legislator does not expect to be pivotal, she should vote with her party leader without the need for leadership threats to avoid paying an electoral cost with no benefit.

However, the political circumstances required to produce Case 3 would be unusual in the modern context, so even if the tools required to apply preference-deviating influence are modest, it won't happen very often. Again, by Assumption 4, we place the party leader to the right of the district median voter. Given that restriction, there are two possible arrays of locations that can produce Case 3, given by Figures 4.3 and 4.4.

Figure 4.3 represents a violation of Assumption 5. While there are some legislators who are more extreme than their party leaders, most legislative sessions contain few roll call votes on which these extreme legislators take positions contrary to their leaders. While this has happened on several occasions since the 2010 election, and chapter 7 will elaborate on the politics of the divided government period from 2011 through 2016, the extremists in those circumstances at least *perceived* the electorally optimal votes as being

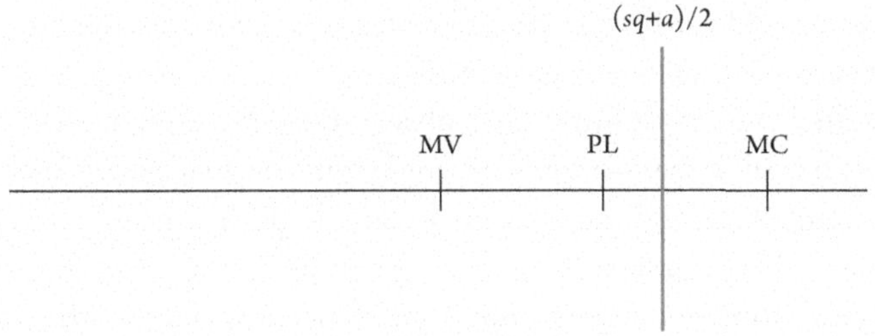

Figure 4.3 Case 3: Preference-deviating influence, variation 3

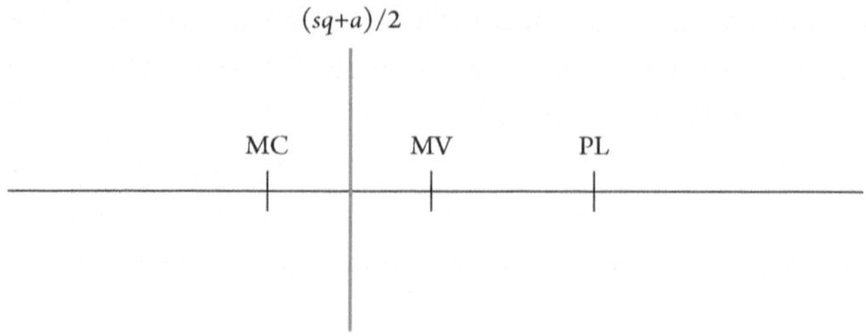

Figure 4.4 Case 3: Preference-deviating influence, variation 4

the *extreme* votes, in which case Figure 4.3 still doesn't apply. While it might seem comparatively easy for the party leader to apply preference-deviating influence with party and electoral influence working together, Figure 4.3 cannot represent normal politics.

Figure 4.4 represents another bizarre circumstance. Given the order of ideal points, the legislator in question is probably closer to the *Democratic* leader's location than to her own leader's location. Historically, of course, there have been analogous circumstances. In the depolarized mid-20th century, there were some northeastern Republicans who were relatively liberal, and southern Democrats who were relatively conservative. In the modern era, though, Figure 4.4 represents a rare circumstance. Thus, while it should be comparatively easy for parties to induce insincere voting when partisan and electoral pressure both work against a legislator's sincere policy preferences, that can't happen very often in the modern era.

However, suppose that a member is more constrained by her primary electorate than her general electorate. If that is the case, then Figure 4.4 might make more sense. Suppose the legislator comes from a heavily slanted district, and the general election is a foregone conclusion. In that case, the general election median voter is not the concern. The primary median voter is. That might create an array of ideal points consistent with Figure 4.4, but that requires a heavily packed congressional district to be represented by someone whose sincere policy preferences are moderate. While there are many marginal districts represented by noncentrists, as we shall discuss in the next chapter, it would be odd for heavily packed districts to be represented by members whose personal preferences are centrist.

The final, and most important, observation about Figure 4.4 is that if it is a circumstance that occurs with any regularity, that means that there are many members of the caucus who are ideologically distant from the party leadership. That implies that a critical condition for party government—ideological homogeneity—is not met. Hence, caucus members would be unwilling to permit the party leader to apply discipline anyway, regardless of the model. While the party leader could apply preference-deviating influence with comparatively small threats in Case 3, Case 3 cannot represent normal politics in the modern era, nor conditions conducive to strong party leadership in any generic model.

Case 4

As we have already discussed, a party leader can use very modest threats to induce a sincere vote because the threat must only counteract the modest electoral damage of a single electorally unpopular roll call vote, without having to counter any policy effects. So, unlike preference-deviating influence, preference-preserving influence never requires harsh disciplinary measures. Furthermore, if it is applied merely to prevent free-riding and solve the collective action problem of sincere voting, then it is not capricious. The remaining question is whether the circumstances in which it might occur are common. Suppose the party leader's preferences are consistent with a rank-and-file member's preferences, but inconsistent with that member's district median voter's preferences. Again, we must place the party leader's location to the right of the district median voter's location. There are two possible arrays of ideal points that can generate Case 4, given by Figures 4.5 and 4.6.

Unlike the figures in Case 3, Figures 4.5 and 4.6 represent conventional legislative politics in a polarized era. There is nothing in either of these

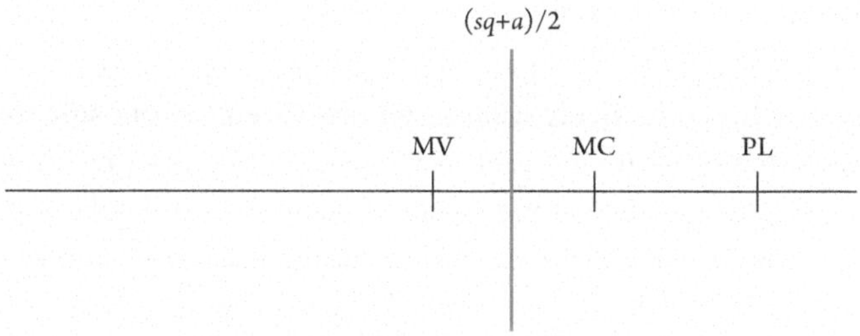

Figure 4.5 Case 4: Preference-preserving influence, variation 1

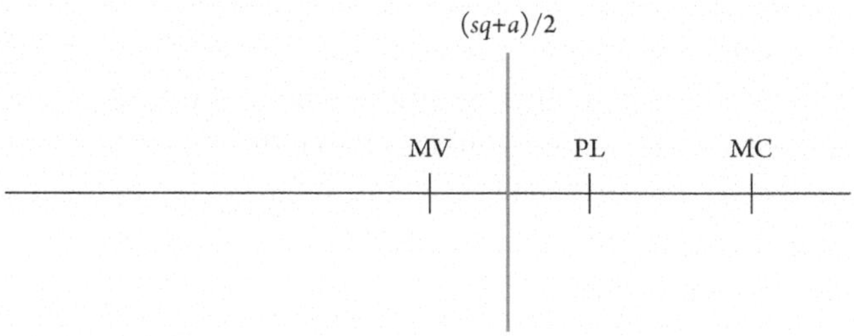

Figure 4.6 Case 4: Preference-preserving influence, variation 2

figures that violates agenda assumptions, and in an era of relatively homogenous parties, we should expect legislators and party leaders to be consistently on the same side of roll call cut points. Furthermore, as we shall see in chapter 5, there are many centrist districts represented by nonmoderates. Thus, preference-preserving influence can be successful with modest disciplinary tools in ordinary circumstances.

Contrasting Preference-Preserving and Preference-Deviating Influence

While we cannot rule out the possibility of preference-deviating influence *ever* occurring in equilibrium, the analysis so far allows us to make some more general claims about the leadership platforms that can exist in equilibrium. When a caucus is ideologically homogeneous, electorally diverse, and policy motivated, that caucus will prefer a leader who solves the collective action problem of sincere voting by applying preference-preserving influence in at least some circumstances to a leader who imposes no party discipline. Moreover, when those same conditions are met, there is a limit on the frequency with which party leaders might *want* to apply preference-deviating influence, and a limit on the frequency with which caucus members would tolerate preference-deviating influence, because the successful application of preference-deviating influence requires either harsh threats, capricious threats, or circumstances that are rare when the conditions for party government are met. The key distinction between the two types of influence is that when the conditions for party government are met, there is a large contingent in the legislative caucus who collectively *want* to be pressured to vote sincerely, and in fact, who need that pressure in order to achieve the policy gains for which they would be willing to sacrifice some electoral security if only given the choice. In contrast, the recipients of preference-deviating pressure never benefit from the pressure applied to them, and when the conditions for party government are met, there cannot be very many of them anyway. Thus, while we cannot rule out the occurrence of preference-deviating influence in equilibrium, it may be more useful to focus on the use of party influence to counteract electoral influence. Parties cannot induce insincere roll call votes very frequently if their leaders want to maintain their positions, and in the modern era, there are few circumstances in which they would *want* to induce insincere votes anyway. However, they can prevent insincere roll call votes, and strategic legislators *want* them to do so when the caucus is ideologically homogeneous, electorally diverse, and policy motivated.

Implications of the Unified Model

If a caucus is ideologically homogeneous, but electorally diverse, then sincere voting on roll calls is a collective action problem. However, if that caucus is policy motivated, it will empower leaders to solve that collective action problem by either threatening modest punishments for insincere votes, or offering modest rewards for legislators who refuse to bow to centrist electoral pressure on roll call votes. When parties act as "preference-preserving institutions," the result will be that caucus members incrementally adopt extreme locations in the policy space, with each extreme roll call vote incrementally reducing their vote shares. However, the electoral damage of each ideologically extreme vote is mitigated by a legislative agenda that includes items that cleanly divide the majority and minority caucuses. Thus, while incumbents will adopt extreme locations that do not maximize their vote shares, they will still generally win re-election, albeit with reduced vote shares, by invoking a paradox that allows extremists to defeat centrists. More importantly, though, incumbents offset the electoral risks they run with policy victories that could not have been achieved without party influence.

This is precisely what we observe at each stage. Members of Congress incrementally adopt ideologically extreme and electorally suboptimal positions in the policy space through their roll call voting patterns. They do so even when they are not pivotal. And they still usually win re-election. The unified spatial model thus explains why incumbents adopt extreme locations and how they win re-election with those locations. By rethinking the nature of party influence in Congress and constructing a spatial model combining roll call voting and legislative elections through a party leadership vote, we can resolve each of the motivating puzzles posed in the introduction.

This model has important implications for competing theories of parties in Congress. "Conditional Party Government" models (Aldrich 1995; Aldrich & Rohde 2000; Rohde 1991) hold that a homogenous legislative caucus will empower party leaders, but critics of the theory argue that a homogenous caucus will vote in a unified manner anyway, making party influence irrelevant. However, that argument comes from ignoring the potential, and indeed necessity, for preference-preserving influence. If parties pressure members to vote sincerely because they are solving the collective action problem of sincere voting, the tautology disappears. Consider the controversial homogeneity condition. A caucus is ideologically homogenous if its members have similar personal preferences, and a caucus is electorally homogenous if its Members represent similar districts. A caucus that is both ideologically and electorally homogenous will vote cohesively regardless of whether or not party leaders apply pressure because even if its members simply deferred to their constituents, the patterns of deference would be uniform. So if the conditions for party government included both ideological and electoral homogeneity, parties would be superfluous, as critics of CPG suggest. However, suppose a caucus is ideologically homogenous but electorally diverse. For its members, sincere voting is a collective action problem. Without parties to solve that problem, members will "free-ride" by voting the way their constituents want, regardless of their own policy preferences. Since their districts have different preferences, this means they will *not* vote cohesively without some external pressure. However, if they attach importance to policy outcomes, relative to electoral security, then they have incentives to empower party leaders to solve the collective action problem of sincere voting, and by doing so, party leaders create cohesion in an ideologically homogeneous caucus that would *not* occur in the absence of party influence, thereby creating policy outcomes that would not occur in the absence of party influence. Hence, the conditions for party government proposed here are ideological homogeneity, electoral *hetero*geneity, and strong policy preferences.[8] By distinguishing between ideological and electoral homogeneity and focusing on party pressure to induce sincere voting, CPG is absolved of a prominent charge against it.

Furthermore, consider how this relates to Krehbiel's (1998) Pivot Point model. That model is largely agnostic as to the origins of legislators' positions in the policy space and simply treats those positions as the net result of the interaction between legislators' various concerns. However, consider empirical analysis of retiring legislators' voting patterns. Such analysis consistently shows that members of Congress vote very similarly before and after they decide to retire (see, for example, Lott 1987; Lott & Reed 1989; Lott & Davis 1992; Bender & Lott 1996; Rothenberg & Sanders 2000). Since they are no longer subject to pressure after they decide to retire,[9] retiring legislators' voting patterns should reflect their sincere personal policy preferences. Since their voting patterns are so similar to their preretirement patterns, this suggests that most roll call voting is sincere all along. This would be puzzling in the context of Groseclose and Milyo (2010), but not in the context of this model. Sincere voting occurs because legislators empower party leaders to incentivize it. Sincere voting is not indicative of parties' irrelevance—it is indicative of their efficacy.

What remains, then, is to demonstrate the empirical value of the unified model. Chapter 5 will examine the changes over time in legislative voting scores as they relate to the conditions for party government, and chapter 6 will discuss how the process of passing healthcare reform in 2010 matches the contours of the model.

5

Polarization and Solving the Collective Action Problem

Chapter 4 derived a model in which an ideologically homogeneous but electorally diverse caucus empowers party leadership to solve the collective action problem of sincere voting, thereby allowing the caucus to achieve policy goals that would otherwise be unattainable, but at the cost of some electoral security. The question now is how this helps us understand the development of polarization in the post–World War II Congress.

Consider the development of polarization in the House of Representatives from the 83rd House to the 111th House.[1] As we shall see, this period saw the development of a collective action problem of sincere voting, creating the conditions under which policy-motivated legislators delegate power to party leadership to solve that problem. While the caucuses have become more ideologically homogeneous, they are as electorally heterogeneous as they were in the mid-20th century. This indicates the development of the collective action problem of sincere voting, which a policy-motivated caucus will solve by empowering party leadership to enforce sincere voting. Consequently, legislators from centrist districts track their party medians over time rather than adhering to their centrist districts' preferences.

Persistent Electoral Heterogeneity and the Collective Action Problem of Sincere Voting

The collective action problem of sincere voting comes from a conflict between legislators' noncentrist personal beliefs and their centrist constituents' beliefs. Consider, then, how these conditions developed over time. Conventional wisdom attributes polarization in Congress to "gerrymandering." According to this argument, state legislatures have increasingly drawn congressional districts in ways that protect incumbents and prevent "competitive" elections. Supposedly, Democratically-held districts are more homogeneously Democratic than they used to be, Republican-held districts are more homogeneously Republican than they used to be, and consequently, incumbents no longer face the risk of a competitive general election. According to the "marginality hypothesis," members of Congress from competitive districts must adopt centrist positions in order to win re-election, so sequential gerrymanders have destroyed legislators' incentives to converge to the absolute center. Hence divergence.

While the marginality hypothesis is controversial (see, for example, Brunell 2006; Fiorina 1973; Sullivan & Uslaner 1978), the underlying premise fits well with chapter 4. Following from Wright et al.'s (1985) state-level analysis, a heavily Democratic district should have a median voter relatively far to the left, and a heavily Republican district should have a median voter relatively far to the right. Gerrymandering, then, has the potential to shift the distribution of district median voters away from the center. If representatives simply cast roll call votes in deference to noncentrist district median voters, then they will incrementally adopt relatively extreme positions in the policy space. Moreover, they will do so without party discipline because electoral pressure incentivizes extreme positioning. Quite simply, the collective action problem of sincere voting wouldn't exist. Legislators with noncentrist preferences would incrementally adopt their noncentrist preferences through their roll call voting patterns with sincere roll call votes. We would observe polarized voting scores without party influence.

The problem is that scholarly analysis has consistently shown that redistricting has had at most a very small effect on legislative polarization (Abramowitz et al. 2006; Brunell 2006, 2008; Buchler 2009, 2010a, 2010b; McCarty et al. 2009). Why? Because competitive districts haven't gone anywhere. Figure 5.1 shows the proportion of House districts in each presidential election from 1952 through 2012 in which the two major party nominees are within either 10 or 20 points of each other in the two-party vote within the district (data generously provided by Gary C. Jacobson).

If we accept Wright et al.'s (1985) argument that presidential vote shares within a polity serve as a useful proxy for the location of its median voter, then the proportion of districts with centrist median voters has not diminished over time. The proportion of districts in which the presidential nominees are separated by less than 10 points has fluctuated more or less randomly around 25%, and the proportion of districts in which the candidates are within 20 points of each other fluctuates around 50%. Conventional wisdom on district lines is that we have sequentially eliminated marginal districts with incumbent protection gerrymanders. In reality, we have just as many competitive districts as in the depolarized era of the 1950s.

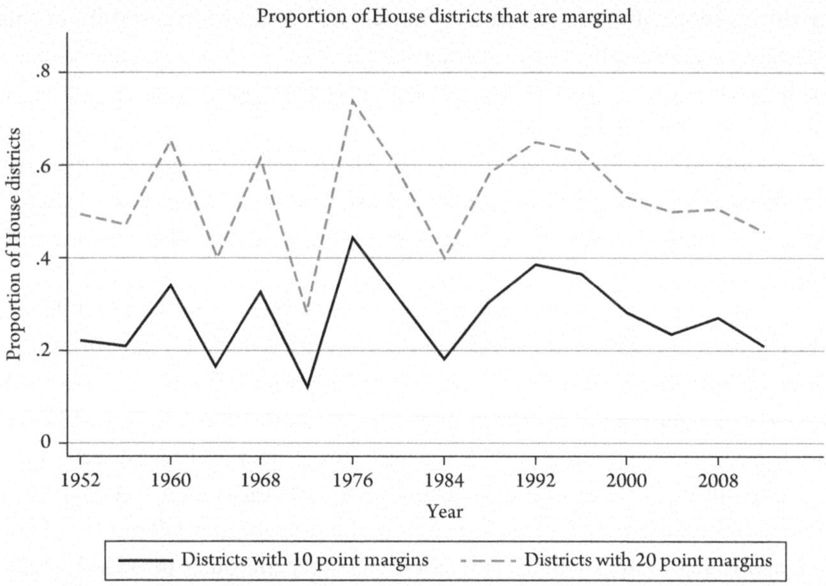

Figure 5.1 Proportion of House districts that are marginal

Consider what this means for incremental positioning. The party caucuses are as electorally heterogeneous as they were in the mid-20th century. If legislators with extreme *personal* preferences replace centrists in such an electorally heterogeneous context, what should happen? Sincere voting becomes a collective action problem, and in the absence of party influence, the legislators who represent centrist *districts* should cast roll call votes as centrists regardless of their personal preferences because, if they aren't pivotal, there is no reason to pay the electoral cost of doing otherwise. So how are the representatives from these centrist districts positioning themselves?

Deference to the Party, or Deference to Constituents?

The party caucuses are as electorally heterogeneous as ever. If representatives with extreme *personal* preferences replace legislators with centrist preferences in an electorally heterogeneous caucus, what should happen? Chapter 4 presented a model in which an ideologically homogeneous but electorally diverse caucus with strong policy preferences empowers party leadership to enforce sincere voting. If they do so, then even legislators from centrist districts should vote like the rest of the caucus because the caucus is ideologically homogeneous and everyone votes sincerely. If they don't empower party leadership in this manner, then the representatives from centrist districts will continue to cast roll call votes as centrists regardless of how extreme their personal preferences are because to do otherwise is to pay an electoral cost with no policy benefit, except in the rare case of casting a pivotal roll call vote.

This suggests a relatively simple approach to analyzing voting scores over time. Do representatives from centrist districts track their party medians as those medians diverge, or do they remain responsive to their own centrist constituents?

For each odd-numbered Congress[2] from the 83rd through the 111th House, and each party, we can regress each legislator's score within the first dimension of DW-NOMINATE on the Democratic presidential candidate's share of the two-party vote within the district from the previous election. These are bivariate models because the point is not to separate direct from indirect effects nor to rule out confounds. The point is to generate a simple measure of overall responsiveness. Note, also, that the total samples within a Congress will not be 435, for two reasons. Estimates of district presidential vote, generously provided by Gary C. Jacobson, do not exist for every district over time, for a variety of reasons. Also, some representatives resign or die, leading to a special election, yielding more than one representative from that district in a given Congress. Each will have NOMINATE scores, and both are included in the analysis because district composition should be essentially the same. Tables 5.1 and 5.2 show the results.

The overall pattern is consistent with Griffin's (2006) argument that district marginality, measured with presidential vote share, predicts centrism at least to some degree, even if a legislator's own vote share doesn't. The coefficients for district Democratic presidential vote are consistently negative and highly statistically significant. Thus, the better the Democratic presidential nominee performs in the district, the more liberal the legislator's NOMINATE score is.

Scale matters, though. Let us consider the predicted DW-NOMINATE scores for representatives from districts that are precisely fifty-fifty in a given election, that is, fitted NOMINATE scores for representatives in whose districts the Democratic presidential candidate got 50% of the two-party vote. Moreover, let us plot these estimates over time alongside the party medians in each Congress. That graph is presented in Figure 5.2.

While the regression results showed that legislators from packed districts incrementally adopt positions that are at least somewhat more extreme than legislators from competitive districts, the pattern in Figure 5.2 puts the difference in context. Over time, legislators from competitive districts incrementally adopt positions that track, not their *district* medians, but their *party* medians. Members from competitive districts are adopting noncentrist, electorally suboptimal locations in the policy space through their roll call voting patterns, and thereby paying electoral costs for being responsive to their party caucuses rather than

Table 5.1 **OLS Estimates: Republican Ideology Scores by Congress**

	83	*85*	*87*	*89*	*91*	*93*	*95*	*97*
Democratic pres. vote	–.0100* (0.0015)	–.0071* (0.0016)	–.0100* (0.0020)	–.0070* (0.0011)	–.0106* (0.0015)	–.0110* (0.0016)	–.0098* (0.0020)	–.0102* (0.0015)
Constant	0.6225 (0.0565)	0.4814 (0.0593)	0.6615 (0.0870)	0.6071 (0.0634)	0.6621 (0.0644)	0.5740 (0.0542)	0.6420 (0.0907)	0.6559 (0.0614)
N	216	201	172	142	194	195	147	193
F	46.39	18.62	25.43	40.64	50.73	47.86	24.51	45.23

	99	***101***	***103***	***105***	***107***	***109***	***111***
Democratic pres. vote	–.0133* (0.0015)	–.0127* (0.0016)	–.0083* (0.0014)	–.0079* (0.0014)	–.0069* (0.0013)	–.0070* (0.0013)	–.0052* (0.0014)
Constant	0.7522 (0.0528)	0.8315 (0.0638)	0.7894 (0.0667)	0.8785 (0.0651)	0.8478 (0.0546)	0.8885 (0.0537)	0.9062 (0.0637)
N	182	178	178	231	227	235	183
F	76.02	64.24	33.45	34.03	29.64	27.71	13.07

DV: First Dimension, DW-NOMINATE.

*$p < .001$, one-tailed probabilities.

Table 5.2 **OLS Estimates: Democratic Ideology Scores by Congress**

	83	*85*	*87*	*89*	*91*	*93*	*95*	*97*
Democratic pres. vote	−.0047* (0.0013)	.0009 (0.0012)	−.0063* (0.0010)	−.0098* (0.0007)	−.0086* (0.0010)	−.0102* (0.0006)	−.0061* (0.0010)	−.0073* (0.0009)
Constant	0.0696 (0.0733)	−0.2664 (0.0638)	0.1295 (0.0594)	0.3491 (0.0429)	0.2023 (0.0566)	0.1177 (0.0260)	0.0521 (0.0564)	0.0849 (0.0455)
N	211	234	267	292	246	245	290	243
F	12.80	0.50	37.63	212.91	75.03	290.79	38.17	70.93

	99	***101***	***103***	***105***	***107***	***109***	***111***
Democratic pres. vote	−.0086* (0.0006)	−.0089* (0.0006)	−.0096* (0.0006)	−.0075* (0.0006)	−.0063* (0.0005)	−.0067* (0.0005)	−.0079* (0.0004)
Constant	0.0959 (0.0303)	0.1548 (0.0309)	0.2506 (0.0358)	0.1324 (0.0369)	0.0290 (0.0304)	0.0396 (0.0317)	0.1552 (0.0286)
N	255	262	258	210	212	202	261
F	184.55	234.00	261.58	177.03	168.34	169.84	307.53

DV: First Dimension, DW-NOMINATE.
*$p < .001$, one-tailed probabilities

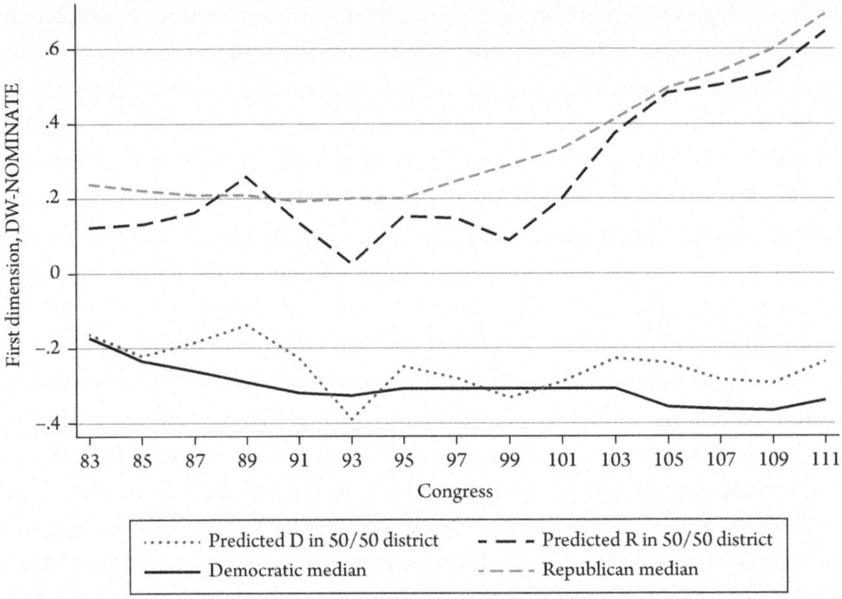

Figure 5.2 Party medians and predicted ideologies of representatives from marginal districts

their district median voters. The party caucuses are voting cohesively despite persistent electoral heterogeneity.

Somebody is solving the collective action problem of sincere voting. Without someone to solve that problem, legislators from marginal districts should incrementally adopt their constituents' centrist positions. They don't. Instead, they track their party caucuses. Is this because they share the policy preferences of their fellow partisans? Undoubtedly so, but since their constituents *don't* share those preferences (recall that we are discussing legislators from marginal districts), that is insufficient to explain the roll call voting patterns we are seeing.

Instead, what we see is the development of the conditions for party government. There are three primary conditions for party government: ideological homogeneity, electoral heterogeneity, and strong policy preferences. The development of ideological homogeneity has been well documented elsewhere. The continued presence of competitive districts, while surprising to some, demonstrates the persistence of electoral heterogeneity. The challenge from a measurement perspective is the strength of policy preferences. Separating ideological noncentrism from strong policy preferences is empirically challenging. One might even pose the notion that the only change that has really occurred is that legislators have placed increasing weight on their policy preferences relative to electoral security. As Groseclose and Milyo (2010) argue, though, that shouldn't matter *without* parties. Where the strength of policy preferences *does* matter is in legislators' willingness to delegate power to party leaders to solve the collective action problem of sincere voting. After all, doing so means sacrificing some electoral security for policy gains. This suggests another possible reason for the development of polarization: legislators placing increasing weight on policy outcomes and decreasing weight on electoral security in their utility functions. In principle, legislators in a less polarized period than today may have been as noncentrist in their *personal* preferences as legislators today, but simply put less weight on those preferences, making them less willing to delegate authority to party leaders. In principle, what we may be observing, then, is the strengthening of policy preferences rather than change over time in their extremism.

From a measurement perspective, this possibility is difficult to assess. We have methods to uncover policy locations, but not preference intensity, and that is at issue here. Of course, given the historical divisions between Southern and non-Southern Democrats through the mid-20th century, and the role of geographic change in producing polarization in Congress, a change in priorities alone cannot be a complete explanation for the development of polarization in the post-World War II period, but it remains worth consideration as a potential component.

Even so, the presence of electoral heterogeneity would make achieving policy goals into a process fraught with a collective action problem that parties are best able to solve. That means we have seen the development of the conditions under which rational caucuses will delegate power to party leaders to solve the collective action problem, and that is precisely what we see. Even legislators from centrist districts defer, not to their constituents, but to their party caucuses. That only makes rational sense if they have delegated power to party leaders to solve the collective action problem of sincere voting, just as the unified model in chapter 4 predicts.

Alternative Solutions to the Collective Action Problem

The fact that legislators from centrist districts so closely track their party medians strongly suggests party influence, but in principle there may be other solutions to the collective action problem. First, consider interest groups. Interest groups have important stakes on many roll call votes and could potentially act as preference-preserving institutions. We know, for example, that interest groups actively contribute to incumbents who share their policy preferences. The conventional scholarly explanation for this pattern is that campaign contributions are simply endogenous, and that they have minimal influence on roll call votes. Ideologically motivated groups simply want to aid the re-election efforts of like-minded legislators.

In principle, though, we might take another interpretation. Suppose that a legislator with noncentrist preferences faces a roll call vote on which the cut point separates her from her district median voter. In the absence of any solution to the collective action problem of sincere voting, the legislator will vote insincerely as though her ideal point is on the same side of the cut point as her district median voter. However, interest groups could offer her a financial incentive to vote sincerely, counteracting the electoral pressure to adopt an incrementally more centrist location. In fact, the campaign contribution might reduce the electoral penalty of an incrementally more extreme bundle of roll call votes anyway. So perhaps interest groups contribute to like-minded legislators and in doing so counteract both the electoral pressure to vote as centrists and the electoral consequences of ideologically extreme voting patterns.

While tempting, that argument is difficult to reconcile with existing econometric analysis. Even if campaign contributions have preference-preserving influence rather than preference-deviating influence, the counterfactual remains the same—in the *absence* of campaign contributions, legislators would cast more centrist roll call votes. Two-stage estimation techniques that account for endogeneity examine that counterfactual. Moreover, as Smith's (1995) review of the collective findings on the influence of campaign contributions suggests, the influence of campaign contributions is neither large nor robust. It is possible that the effects of campaign contributions vary based on issue salience and a legislator's electoral circumstances (Buchler 2011b), but even those estimated effects are modest in the larger picture. Given the inconsistent evidence for the influence of campaign contributions, we cannot attribute the broad trends toward polarization to interest group influence, preference-preserving or otherwise. Interest groups could *help* solve the collective action problem of sincere voting and thereby contribute somewhat to polarization, but the fact that legislators from centrist districts so closely track their party medians makes party influence the simpler explanation.

In the current political environment, though, there may be an even more tempting explanation than interest group influence—the influence of primaries. Perhaps the political system has shifted from one in which incumbents fear losing a general election to one in which they fear losing a primary. The structure of the electoral system makes the explanation at least plausible. After all, elections are sequential processes in which the parties nominate their candidates, who then face each other in the general election. If a candidate loses a primary, that candidate doesn't even make it to the general election.[3] Moreover, if primary voters are more ideologically extreme than general election voters, then the need to win a primary first might force incumbents to incrementally adopt more extreme positions than they would if they only had to run in the general election. Empirically, we do observe lower vote shares for incumbents who seem "out of step" with their primary constituencies (Brady et al. 2007), so perhaps rather than parties solving the collective action problem *within* Congress, primary electorates are solving the collective action problem by changing the direction of the electoral pressure.

Primaries have existed since their introduction as Progressive Era reforms, but that does not mean that their effects must be constant over time. After all, it is not necessary for primary voters to insist on ideological purity. A depolarized era might occur because, in that era, primary voters strategically vote for centrists in order to nominate the most electable general election candidate. A polarized era might result, then, from a shift toward sincere voting in primaries. That would put increasing electoral pressure on incumbents to adopt more extreme positions given the structure of the electoral process. So is Congress polarized simply because legislators used to fear general elections, but now fear primaries, making party influence within Congress unnecessary to solve the collective action problem of sincere voting?

Consider two modern developments. First, the movement toward extremism has been more dramatic among Republicans than among Democrats, and, second, conventional wisdom is that Republicans now fear losing a primary if they are insufficiently conservative more than they fear losing a general election for being too extreme. Perhaps polarization among *Republicans*, at least, results not from pressure within the legislature, but from increasing fear of losing a primary.

Tempting though this story is, primaries have limited explanatory power, both in the timing and in the magnitude. Republican fears about primary challenges became most evident in 2010 with the rise of the "Tea Party." The Tea Party began as a protest movement whose participants were largely Republican (see, for example, Skocpol & Williamson 2012), and it developed a complex relationship with the Republican Party. Some Republicans who identified more with the Tea Party than the Republican Party, like former Senator Jim DeMint (SC), tried to harness the Tea Party to purge moderates from the Republican congressional ranks. In a speech at the 2010 meeting of CPAC (the Conservative Political Action Conference), DeMint said, "Let me make myself even clearer: I'd rather have 30 Marco Rubios in the Senate than 60 Arlen Specters." Rubio was, at the time, a candidate for Florida's open Senate seat and a favorite among conservative activists.[4] Specter was in the final year of his Senate career. For most of that career, he was a Republican, although a relatively moderate one who frequently clashed with party leaders, as well as the more conservative members of the caucus. In 2009, he switched parties, and chose to run as a Democrat because he believed that he stood a better chance of winning a Democratic primary in 2010 than he would have stood in a rematch against his 2004 primary opponent, former Representative Pat Toomey (R-PA).

For DeMint, ideological purity was the sine qua non of the Republican Party. With only 30 senators, no matter how conservative, Republicans could not even successfully filibuster Democratic legislation. With 60, though, even if those 60 were moderates, they could defeat Democratic filibusters and pass legislation that would be considerably further to the right than what would be passed by a Senate consisting of 70 senators with preferences similar to Senator Bernie Sanders (I-VT). However, DeMint claimed that he would sacrifice that for the sake of ideological purity within the party, believing that to serve his long-term policy goals more effectively. And he was not the only one who felt that way. In the 2010 primaries, the Republican Party nominated several Senate candidates who were seen as ideologically pure but, for various reasons, could not win seats that otherwise should have been easy Republican victories in a midterm election with an unpopular Democratic president and a weak economy. The most obvious case was Christine O'Donnell. In a surprise upset, O'Donnell defeated Representative Mike Castle for the Republican nomination to Delaware's open Senate seat. Prior to O'Donnell's victory, the seat was considered a Republican lock because of Castle's popularity in the state and timing that would favor Republicans. However, Castle's reputation as a moderate made him anathema to the Tea Party, and he lost to a perennial candidate with no elected experience. O'Donnell became known nationally more for her gaffes than anything else. In a television appearance years earlier, she claimed that as a youth, she had "dabbled into [rather than 'in'] witchcraft." During the 2010 campaign, her most famous ad referenced the clip, and began with the now infamous line, "I am not a witch." Unsurprisingly, O'Donnell lost the general election.

In Nevada, Senate Majority Leader Harry Reid was considered a prime target for Republicans. He was widely unpopular in the state, and all Republicans needed to do was to nominate someone plausible to defeat him. Instead, they nominated Sharron Angle. Unlike O'Donnell, Angle had elected experience as a member of the Nevada State Senate. However, she shared other important characteristics with O'Donnell. Most importantly, Angle's conservatism made her popular among the Tea Party but unpalatable to most others. In her most famous statement, she said, "I hope that's not where we're going, but you know if this Congress keeps going the way it is, people are really looking toward those Second Amendment remedies and saying, my goodness, what can we do to turn this country around? I'll tell you the first thing we need to do is take Harry Reid out." Angle repeatedly refused to either retract or even clarify the statement, and voters in Nevada decided that however much they despised Harry Reid, they could not replace him with Sharron Angle.

O'Donnell and Angle were not the only Tea Party primary victories. They simply stand out because they lost seats that should have been easy Republican victories in 2010. Other Tea Party insurgents actually won their general elections.

In Kentucky, Senate Minority Leader Mitch McConnell's (R-KY) choice to replace retiring Senator Jim Bunning was Trey Greyson, who lost the primary to Tea Party favorite and 2010 general election winner Rand Paul. In Florida, Governor Charlie Crist chose not to face the aforementioned Marco Rubio in the Republican primary for the open Senate seat, so he ran as an independent and lost the general election to Rubio. So Tea Party candidates had a mixed record of success in 2010 general elections. The important point, though, is what their successes in *primaries* may have done to the mentality of Republican legislators. There were enough high-profile races in which a Tea Party–identified candidate defeated a more moderate candidate that Republicans across the country started to become very nervous that if they were insufficiently conservative, they would lose a primary to a Tea Party candidate. This put non–Tea Party Republicans in a bind. The Republican Party stood to benefit from increased turnout driven by Tea Party activism, but that very same activism posed a threat to Republicans who failed their ideological purity requirements in a primary, and those same purity requirements made the Republicans who met them weaker in the general election. Nevertheless, the primary precedes the general, so perhaps the need to fend off a primary challenge forces legislators to incrementally adopt more extreme locations than they would if they only had to run in general elections. If so, then we can explain polarization without parties solving the collective action problem of sincere voting within the legislature.

Empirically, though, the attention recently paid to the threat of a primary challenge in the Republican Party is puzzling because so few Republican incumbents have actually lost for being insufficiently conservative. The Tea Party victories in 2010 were almost all in races without an incumbent, such as Rand Paul's victory in Kentucky and Marco Rubio's victory in Florida. Even O'Donnell and Angle didn't defeat incumbents for their nominations. O'Donnell defeated Castle for the nomination to an open Senate seat, and Angle defeated Sue Lowden for the nomination to challenge a *Democratic* incumbent. In 2010, only four Republican incumbents failed to secure renomination: Representative Bob Inglis from South Carolina, Senator Lisa Murkowski from Alaska, Representative Parker Griffith from Alabama, and Senator Robert Bennett from Utah. None fit the popular storyline. Inglis comes closest. Inglis lost his primary in 2010, and hence his seat, for no discernible reason other than a lack of support among his primary constituency because of a *perception* that he was too moderate. This was likely more dispositional than anything else, given his DW-NOMINATE score in the 111th House of .8930. His most famous instance of being at odds with the Tea Party was during a meeting with constituents in 2009 at which he told attendees to "turn Glenn Beck off." Beck was, at the time, the host of a show on Fox News that was popular among Tea Party activists, and distancing himself from Beck did not endear Inglis to the many fans of Beck among conservative activists. While there is little direct evidence that this comment cost Inglis the primary in 2010, he may have lost because he was *seen* as insufficiently conservative, despite his voting record indicating otherwise. At the very least, there are no better explanations. However, Inglis is the only incumbent from 2010 who even comes close to fitting this story. Consider the other three cases from 2010.

Senator Murkowski lost a primary to Joe Miller, a self-identified Tea Party candidate, who accused her of being insufficiently conservative. In fact, Murkowski was a moderate. However, after losing the primary, she ran in the general election as an independent candidate. Her name didn't even appear on the ballot, though, forcing her to run a write-in campaign. Nevertheless, she won. She retained her seat despite the primary loss, just as Senator Joe Lieberman (D-CT) had done in 2006 after losing a primary to Ned Lamont because of his support for the war in Iraq. The ability to survive a primary loss in this manner should strengthen, not weaken, legislators' resolve to resist primary pressure toward extremism. No such option exists for incumbents who lose a general election.

Of course, not everyone who loses a primary can use the Murkowski-Lieberman gambit. Consider Representative Griffith. Griffith's loss in his 2010 primary should have been no surprise. He won his first election to the House of Representatives in 2008 as a Democrat. He switched parties midway through his first term, fearing that 2010 would be a bad year for Democrats. However, Republican primary voters were unconvinced and declined to renominate, or rather nominate, him. This was not simply a matter of insufficient conservatism—it was distrust for a party switcher. Senator Arlen Specter of Pennsylvania faced a similar problem in 2010. He switched from the Republican Party to the Democratic Party in 2009 because he believed that if he ran for re-election as a Republican, he would lose a primary to former Representative Pat Toomey, whom he narrowly defeated in the 2004 Republican primary. Democratic primary voters were unconvinced by Specter's switch and chose instead to nominate Representative Joe Sestak, although Sestak went on to lose to Toomey in the 2010 general election anyway. Specter may serve as a warning to Republicans, having been essentially chased out of the party for being too moderate, but his party switch both muddies the waters on that matter, because he didn't actually lose to Toomey, and demonstrates why Griffith's 2010 loss does *not* serve as an indicator about what happens to insufficiently extreme legislators. Party switchers may just have difficulty winning the primary in the party that is used to voting against them. So neither Murkowski nor Griffith serves as a clear warning to insufficiently conservative Republican legislators.

Of course, one of the most commonly cited cautionary tales from 2010 is even stranger. Senator Bennett is frequently cited as an example of someone defeated for his infidelity to conservatism, but he did not actually lose a primary. The Utah Republican Party had an odd nominating system that required candidates to have their names placed on the *primary* ballot by a nominating convention. Party delegates refused to place his name on the primary ballot. While this was a punishment for his lack of conservative purity, and in particular his willingness to work on a healthcare proposal with Senator Ron Wyden, a liberal Democrat from Oregon, it says nothing about whether or not insufficient conservatism will lead to a primary loss because Bennett didn't even get to the primary. Instead, it indicates something about the need for conservative purity among Republicans representing Utah. Of course, Utah's other senator, Orrin Hatch, took the lesson to heart and abandoned any efforts at bipartisanship to avoid a similar fate in 2012. Prior to the Bennett incident, Hatch regularly trumpeted not just his conservatism, but his record on bipartisanship, including his relationship with Senator Ted Kennedy before his death. After Bennett, Hatch kept quiet about Kennedy and toed the rhetorical party line. In 2012, his party allowed his name to remain on the primary ballot. While Bennett served as an important cautionary tale, it was mostly of importance to Orrin Hatch since Bennett didn't lose a primary. He was simply the casualty of Utah's unique nominating system.

So of the four Republican incumbents who failed to secure renomination in 2010, at best only one—Representative Bob Inglis of South Carolina—lost his seat because he was defeated in a primary for being insufficiently conservative, and even that was dispositional rather than anything having to do with his voting record.

In 2012, seven Republican incumbents were denied renomination: Senator Dick Lugar from Indiana, Representative Don Manzullo from Illinois, Representative Ben Quayle from Arizona, Representative Sandy Adams from Florida, Representative Cliff Stearns from Florida, Representative Jean Schmidt from Ohio, and Representative John Sullivan from Oklahoma. Only Lugar fits the popular storyline. In 2012, with a DW-NOMINATE score of .2280, he lost a primary to Richard Mourdock—a Tea Party challenger who expressed his attitude toward bipartisanship as follows: "I certainly think bipartisanship ought to consist of Democrats coming to the Republican point of view." Lugar lost because he was seen as insufficiently conservative. Like O'Donnell and Angle, though, Mourdock went on to lose the general election to Representative Joe Donnelly, a Democrat, after describing a pregnancy resulting from rape as "something God intended."

However, none of the other six Republicans who were denied renomination should serve as warnings to Republicans who want to avoid the label of "RINO" (Republican in name only). Representative Manzullo lost a primary, not because of a Tea Party challenge, but because redistricting placed him in the same district as fellow Republican incumbent Adam Kinzinger. The same happened to Representative Quayle, who wound up in a primary against Representative David Schweikert, and Representative Adams, who wound up in a primary against Representative John Mica. When redistricting forces two incumbents to face each other in a primary, somebody has to lose, but the results have limited implications for any incumbent *not* forced into such a match.

Representative Stearns was not forced into a member-to-member primary, but his district was heavily redrawn, and with a DW-NOMINATE score from the 112th House of .8450, he could not be accused of insufficient conservatism.

Representative Schmidt's loss belies easy explanation, but insufficient conservatism seems a stretch. Her DW-NOMINATE score of .7230 was almost exactly the party median in the historically extreme Republican caucus of the 112th House of Representatives, and her reputation was as a firebrand to boot. As a first-term Representative in 2005, Schmidt made her first big splash debating the continuation of the Iraq war with a highly decorated marine veteran, Representative John Murtha (D-PA). She said, "Cowards cut and run, marines never do." A Republican willing to call Murtha a coward was not exactly a Bob Bennett or a Dick Lugar. Whatever caused Schmidt to lose, it was neither moderation nor accommodation.

Representative Sullivan's loss was similarly odd. He did not have Schmidt's reputation as a firebrand, but he was no moderate. His DW-NOMINATE score in the 112th Congress was .7260. Perhaps a lack of attention to the campaign cost him the primary,[5] but the ideological purge storyline just doesn't fit. Once again, the only incumbent primary loss from 2012 that fits the Republican purge storyline is that of Senator Dick Lugar.

The 2014 round of primaries was somewhat more complicated. Only two Republican incumbents were denied renomination. Representative Ralph Hall from Texas faced a loss complicated by his decision to make 2014 his last campaign, given his advanced age. The more interesting story was the surprise loss of former Majority Leader Eric Cantor, from Virginia, to David Brat in a contest that nobody expected Cantor to lose. Early polling showed him in no danger, and Brat had no money to mount a vigorous campaign, yet Cantor lost. The common explanation was that Cantor lost, not for any general lack of conservatism, but for his openness to immigration reforms that Tea Party activists opposed on the grounds that they constituted "amnesty" for undocumented immigrants. The biggest irony may be that Cantor had positioned himself as a sort of Tea Party–friendly Speaker-in-waiting if Boehner compromised too much with President Obama and Senate Democrats. Nevertheless, Cantor's surprise loss may serve as a strong motivator for other Republicans not to compromise on immigration, even if that loss was issue-specific.

In 2016, three more Republicans lost primaries: Renee Ellmers (NC), Randy Forbes (VA), and Tim Huelskamp (KS). Ellmers's DW-NOMINATE score in the 113th House was .5870 and Forbes's score was .6290, putting them somewhat to the left of the party median, but only because the party median at the time was .735. More importantly for Ellmers, North Carolina had just redrawn its districts mid-decade, putting her in a precarious position, causing her to lose narrowly—hardly a true cautionary tale for moderates.

Then there's Tim Huelskamp, whose DW-NOMINATE score in the 113th House was .9960. He was often at odds with Speaker Boehner because he believed that the party "establishment" wasn't conservative *enough*. Chapter 7 will elaborate on his relationship with Speaker Boehner, but in 2016, Huelskamp faced a challenge from an "establishment"-backed candidate named Roger Marshall. Marshall won. 2016 does not contain any true cautionary tales for moderate Republicans worried about a primary challenge. Rather, it contains a cautionary tale that intransigent *extremists* like Huelskamp are the ones in danger of losing a primary.

In all of 2010 and 2012, then, only one Republican incumbent clearly lost his seat because of a Tea Party primary challenge for being insufficiently conservative in how he positioned himself through his roll call voting: Lugar. Inglis's problem was dispositional, and the big shock of Cantor's 2014 loss has been interpreted as being about immigration rather than general conservatism. 2016 contained no clear cautionary tales for moderates. Rather, it contained a cautionary tale for extreme conservatives in the form of Huelskamp's loss. If Republican incumbents are now terrified of losing a primary, that would seem to be an overreaction. However exaggerated, though, the *fear* of a Tea Party challenge appears real among many Republican legislators. Moreover, that fear could persuade legislators to adopt more extreme positions even if it is based more on a few anecdotes than actual trends. In a perfect demonstration of historical irony, this sentiment was best expressed by the sort-of 2010 Tea Party casualty, Bob Inglis, about the true 2012 Tea Party casualty, Dick Lugar: "The strategy is to have an object lesson, to make Dick Lugar an object lesson, and all you have to do is take down one Dick Lugar and you whip 20 into line, because they don't want to be the next Dick Lugar."[6] It is plausible, then, that Republicans *post*-2010 adopt ideologically extreme positions thinking that it is electorally optimal, even if that belief is an overreaction to a few anecdotes. Party discipline may be unnecessary to promote extremism in voting scores, then.

Even if that is true, though, primary pressure is not a compelling explanation for Figure 5.2. First, notice the timing. The quantitative analysis earlier in this chapter conspicuously ended with the 111th Congress—the last Congress *prior* to anecdotes like Inglis and Lugar. It is quite plausible that fear of primaries has changed the dynamic dramatically in Congress since the 2010 election, and chapter 7 will draw out the implications of that much more directly. However, we cannot explain trends toward polarization *prior* to 2010 with a fear of primaries that has developed *in the wake of* 2010.

As Boatright (2013) has observed, the threat of a primary challenge emerges periodically in response to historical circumstances, and, as Abramowitz (2008) and Hirano et al. (2010) argued prior to the Tea Party era, primaries had limited explanatory power with respect to extremism. Republicans may *now* fear primaries more than general elections, but that can't explain a consistent trend that began in the 1970s, which is when the Republican Party started moving rightward.

Moreover, Democrats are also incrementally adopting electorally suboptimal positions through their roll call voting patterns. True, their positions are *less* extreme than Republicans', but there is no compelling evidence that Democrats have begun fearing primaries more than general elections, particularly in the wake of Lieberman's 2006 end run around his primary loss.

Chapter 7 will argue that fear of primary challenges among Republicans *since* 2010 has fundamentally altered the nature of both electoral pressure and party pressure, but as an explanation for the broad trend toward polarization in the post–World War II period, primaries simply will not suffice. Legislators, even from centrist districts, track their party medians over time rather than adhering to their constituents' centrist preferences. This is what we would expect if parties apply preference-preserving influence to solve the collective action problem of sincere voting among an ideologically homogeneous but electorally diverse and policy-motivated caucus. In short, ideological polarization is a manifestation of party influence on roll call votes.

Extensions and Further Testing

The central hypotheses derived from the unified model involve the circumstances in which legislators vote either sincerely or insincerely, and how parties influence that decision. The reason the empirical analysis has been indirect, then, is that directly measuring sincerity is challenging. Legislators will never admit to voting in ways that don't reflect their views of good policy. The basic challenge, then, is that as Smith (2007) has noted, party influence operates in a way that is intrinsically unsuited to direct testing.

Nevertheless, there is at least the possibility of more direct tests. The key is the trade-off between legislators' policy preferences on legislation and voter preferences. While roll call cut points are easy to uncover with NOMINATE, though, the locations of the status quo and alternative points are what really matter, and those are more difficult. Richman (2011) has proposed a method to uncover those, and if combined with the proper polling data, one could construct more direct tests.

Alternatively, one might examine the processes that follow a policy failure. If party organization exists to solve a collective action problem, then we might expect parties to be strengthened in response to specific failures to solve the collective action problem. For example, if the majority party gets rolled on an important piece of majority-sponsored legislation, the caucus might respond by centralizing power in order to prevent future legislation from being stopped by caucus members who are afraid to support the party, along the lines of how Democrats instituted the 1970s reforms in order to prevent southern committee chairs from continuing to block their legislation (Rohde 1991).

While direct tests of party influence are difficult, as Smith (2007) argued, that does not mean we should abandon looking for them.

Reconsidering the Retirement Effect and the Role of Sincerity

The strong suggestion of party influence on roll call voting places existing findings on the sincerity of roll call votes in a new light. Several references have been made so far to the observation that members of Congress seem to vote in ways that reflect their sincere, personal policy preferences. When legislators decide to retire, they are no longer subject to electoral pressure, so their voting patterns after they decide to retire should indicate their personal beliefs. Since voting patterns only change very modestly *after* retirement decisions, legislators' votes are mostly sincere all along. However, when legislators decide to retire, "the electoral connection" is not the only thing that is "severed," to use Rothenberg and Sanders's phrasing. So is party pressure.[7] If party pressure exists merely to counteract electoral pressure, though, the disappearance of electoral pressure eliminates any need for party pressure. When members of Congress who belong to an ideologically homogeneous but electorally diverse and policy-motivated caucus are still running for re-election, they are pulled toward the center by electoral pressure, and back toward their own ideal points by party pressure. When they decide to retire, *both* forces stop acting on them. Electoral pressure disappears because they are no longer running for re-election, and party pressure disappears because that pressure only exists to counteract the now-gone electoral pressure. The result is that retiring legislators remain close to where they have always been—their personal ideal points. Hence, the relative paucity of the retirement effect makes sense in the context of this model. Parties create polarization by preventing legislators from deferring to their moderate constituents. Hence, even the legislators from centrist districts track their party medians rather than their constituents, and remain in the same locations after they decide to retire because those locations were relatively sincere all along. If party pressure merely counterbalances electoral pressure, then it disappears when electoral pressure does, and nothing changes as far as where legislators position themselves when they decide to retire.

Party Influence, Polarization, and the Collective Action Problem of Sincere Voting

The post–World War II period in Congress saw the development of the key conditions for party government derived in chapter 4: ideological homogeneity and the persistence of electoral heterogeneity. That created incentives for a policy-motivated caucus to empower party leaders to solve the collective action problem of sincere voting. With parties enforcing sincere voting among an electorally heterogeneous caucus, we see legislators from centrist districts tracking their party medians rather than their district preferences. If parties were not solving the collective action problem, legislators from centrist districts would adhere to their constituents' preferences rather than their party medians. They would incrementally adopt centrist positions through their roll call votes.

Moreover, while other actors are at work here, party influence remains the most compelling explanation. Interest groups are too weak to create such dramatic effects, and primary pressure is too historically idiosyncratic for such consistent effects. Indeed, the patterns long predate the rise of Tea Party challenges among congressional Republicans, and the effect crosses party lines.

It is always possible that many small factors are at work. Throughout the post–World War II period, we have also seen dramatic changes in the structure of the media; the nature of campaigns; the relationship between parties, interest groups, and think tanks; geographic and demographic trends; and more. Nevertheless, the collective action problem of sincere voting is a mathematical necessity, and the existence of polarization means that it is being solved. The basic fact that legislators from centrist districts track their party medians so closely and so consistently over time suggests that the ultimate answer lies in parties, which are the institutions best able to solve the collective action problem of sincere voting, and whose caucuses require that.

While the overall trends in polarization strongly suggest that parties are functioning as preference-preserving institutions, the next question is whether or not major policy decisions reflect the model. Chapter 6 will examine the passage of the Patient Protection and Affordable Care Act in 2010 and show that it could not have passed without the Democratic Party solving the collective action problem of sincere voting by applying preference-preserving influence.

6

The Collective Action Problem in Practice

While voting scores in Congress strongly suggest that parties are solving the collective action problem of sincere voting, the next question is whether or not we can observe preference-preserving influence at work on substantive political processes. This chapter presents a case study in the operation of Congress when the majority party faces a collective action problem, pitting legislators' personal policy preferences against their electoral incentives, and forcing parties to step into the breach to solve the problem if the caucus is to achieve its noncentrist policy goals.

The House of Representatives and the Patient Protection and Affordable Care Act of 2010

Chapter 3 addressed the Senate vote on cloture for the PPACA because the cloture motion received precisely 60 votes. Thus, it presented an opportunity to study how legislators behave when casting a pivotal vote on which policy preferences and electoral pressure conflict. Here let us turn to the House vote in order to see what happens when there is a collective action problem because nobody is pivotal. The final vote to pass the PPACA in the House of Representatives occurred on March 21, 2010. It was decided by a tally of 219 to 212, so no individual legislator cast a pivotal vote. Given the unpopularity of the bill among most constituencies, any *one* legislator could have cast the electorally optimal no vote—electorally optimal in any district that wasn't sufficiently packed with Democrats, anyway—without changing the outcome, and any *one* legislator from such a district not doing so was paying an electoral cost for no policy benefit. How do we explain this?

Prior to the final House vote on March 21, the House and Senate had passed different versions of the PPACA. Under normal circumstances, that isn't a problem. The House and Senate would appoint members of a conference committee to reconcile the two versions of the bill, which would return to each chamber for a final vote on the conference report. However, between the time that the Senate passed its version of the bill and March 21, the composition of the Senate had changed. When Senator Ted Kennedy (D-MA) died, Massachusetts held a special election to replace him, and surprising many, that special election was won by the Republican candidate, Scott Brown. Brown had campaigned on a promise to vote against any healthcare bill, going so far as to sign his name as "Scott Brown, 41," indicating that he would be the 41st vote against cloture for any reconciled version of the PPACA. Since the Democrats could no longer invoke cloture in the Senate after Brown's victory, they could not reconcile the House and Senate versions of the bill because they couldn't force a floor vote on the reconciled version. The only way for the Democrats to pass a bill would be for the House of Representatives to pass the Senate's version of the bill. As long as the House passed the Senate's version without amendment, there would be no need to reconcile two versions of the bill, and hence no need to invoke cloture again. Both chambers could then use the "budget reconciliation" procedure to pass a set of amendments in a second bill because budget reconciliation bills cannot be filibustered. The limitation on the procedure, known as the Byrd rule (in honor of Senator Robert Byrd from West Virginia), is that only provisions directly related to the budget are immune to filibusters.

The challenges were twofold. First, the Senate version included several provisions that were difficult for the House to accept, such as benefits for Nebraska and Louisiana, known colloquially as "the cornhusker kickback" and "the Louisiana purchase," intended to secure cloture votes from Democratic Senators Ben Nelson and Mary Landrieu. While these provisions could be removed through a budget reconciliation bill after the House passed the Senate's version, Democrats in the House still had to vote for a bill containing these controversial provisions, which nobody wanted to do. However, that was merely a procedural discomfort since a subsequent budget reconciliation bill could repeal these provisions. The greater challenge was that the Senate's version of the PPACA had weaker restrictions on abortion than the House's version, and a set of as many as 12 antiabortion Democrats,[1] led by Bart Stupak of Michigan, indicated that they would not vote for any version of the bill that didn't have the House version's abortion restrictions.

Therein lies the problem. Speaker Nancy Pelosi could not pass the bill without the support of Stupak and the other antiabortion Democrats. Stupak insisted on a set of amendments to the Senate's version as a condition for his support. Yet if the House passed those amendments, the bill would no longer be identical to the Senate's version. In order for the Senate to pass the PPACA with Stupak's amendments, the Democrats would have needed 60 votes to invoke cloture *again*, but with the loss of Kennedy's seat to Scott Brown, they only had 59 votes for cloture. Moreover, since Stupak's amendments were not budgetary, they could not be included in the subsequent budget reconciliation bill. So either Stupak had to back down or healthcare reform was dead. Democratic leaders, then, began to apply pressure to Stupak and the other holdouts. How can we characterize this pressure and its consequences?

First, the electorally optimal vote for Stupak and the other antiabortion Democrats was clearly to vote against the PPACA. Most obviously, the bill did not have strong public support. Nationwide public opinion polls preceding the final vote ran consistently against the bill. Moreover, Stupak's district wasn't exactly a Democratic stronghold. Barack Obama only carried 51% of the two-party vote in that district in 2008, and George W. Bush carried 54% of the vote there in 2004. Since the district was slightly less Democratic than the country as a whole, the bill was probably even less popular there than nationwide. Moreover, Stupak's own actions magnified the danger of supporting the PPACA. He had built an image for himself as a moderate, "pro-life" Democrat, and indicated that he would not support any version of the PPACA without the abortion restrictions he secured in the House's initial version of the bill. That promise itself created the potential for an additional electoral penalty for Stupak if he voted for the Senate's version by opening himself up to the charge of being a flip-flopper, or at least weak. So through district composition and his own actions, the electorally optimal choice for Stupak was clear—vote against the Senate's version of the bill. Moreover, when the final vote happened in the House on March 21, 2010, the tally was 219 in favor and 212 opposed. If Stupak *alone* (this is key) had voted no, the bill still would have passed. He could vote, without any policy consequence, for the electorally optimal choice. He chose not to do so. In the end, he yielded to Speaker Pelosi and President Obama and voted for the Senate's version of the bill. He then decided not to seek re-election in 2010, fearing that his vote on the PPACA would cost him his seat. According to Nyhan et al.'s (2012) analysis, he may well have been right. The next question, then, is whether Stupak voted sincerely or insincerely.

The central issue for Stupak was compliance with the Hyde Amendment. That provision, passed in 1976 and named for former Representative Henry Hyde (R-IL), prohibited the use of federal funds for abortion. There were two abortion questions associated with the PPACA. The first had to do with the subsidies to help low-income individuals purchase health insurance. Both the Senate and the House versions of the bill prohibited recipients of that subsidy from using the subsidy to purchase an insurance plan that included abortion coverage. Individuals receiving the subsidy could only purchase abortion coverage separately using their own money. However, the House version of the bill went beyond that. The PPACA created a set of market exchanges, overseen by either the states or the federal government, in which individuals who did not receive employer-sponsored insurance could compare plans. Overseeing these exchanges required the government to incur some cost. Stupak wanted a provision that prohibited any health insurance company from listing any plan on these exchanges that covered abortion, regardless of whether or not those purchasing the plan were receiving subsidies. The House version had such a restriction. The Senate version didn't.

How should we expect Stupak and his colleagues to view the policy issues at hand? Obviously, Stupak would have preferred a bill that included these additional restrictions to one that did not. So he would clearly have preferred the House version of the bill to the Senate version. However, from a practical standpoint, that didn't matter. The House version could never pass the Senate because without Kennedy's vote, Democrats could never invoke cloture. Moreover, Stupak's amendments could not be included in a subsequent budget reconciliation bill because such provisions were not budgetary, so they had no protection from the filibuster under the Byrd rule. The question was whether Stupak would have preferred the pre-ACA status quo to the Senate version of the bill, or the Senate version of the bill to the status quo.

The Senate bill contained several provisions that were core objectives of the Democratic Party for decades, the most important of which were subsidies to help provide health insurance to low-income individuals, and restrictions that prevented insurance companies from denying coverage to those with "preexisting conditions." Stupak certainly wanted those provisions. The question was whether or not the insurance exchanges would be so odious to Stupak if they didn't have his abortion restrictions that he would have preferred to see the whole bill defeated rather than to create those exchanges. That seems unlikely. Creating the exchanges without Stupak's abortion restrictions would not result in significantly more abortions than the law as it existed prior to the PPACA. The bill would not remove any legal restrictions on abortion, nor would it provide funding for abortion. With or without the bill, those looking for an individual plan that covered abortion could buy one as long as it was with their own money. The only difference would be whether or not they could do so in an exchange that would slightly reduce the time and effort required to find the plan. So passing the PPACA without Stupak's abortion restrictions would have moved abortion policy a very small distance away from Stupak's abortion preference.

However, the subsidies to help low-income individuals purchase insurance and the prohibition on companies denying coverage to those with preexisting conditions were centerpieces of the Democratic Party's agenda. Those provisions were among the most significant policy changes Congress had considered in decades, moving policy dramatically toward the Democrats' policy preferences on healthcare.

Given that contrast between the magnitude of the changes to abortion policy and the magnitude of the changes to nonabortion policy, perhaps we should think of Stupak's preferences in multidimensional terms.[2] The Senate's version of the PPACA must have been closer to Stupak's preferences than the status quo with respect to *non*-abortion-related issues, and with respect to abortion, it was at most very modestly further from his preferences than the status quo. Thus, it is difficult to believe that he saw the Senate's version of the PPACA as further from his overall ideal point in Euclidean terms than the status quo. The most viable interpretation of his behavior, then, is that Stupak personally preferred the Senate's version of the PPACA to the status quo, and his statements throughout the process were bargaining tactics in order to extract as many concessions as possible, which worked until Senator Kennedy's death and eventual replacement by Scott Brown.

This interpretation received further support from the eventual resolution. In the end, Stupak agreed to vote for the Senate version in exchange for an executive order, issued alongside President Obama's signature on the final PPACA, repeating that the Hyde Amendment remained in effect. This was a symbolic concession to Stupak in an attempt to give him a face-saving way to back down from the statements that he had made as bargaining tactics when he ran out of leverage.

The critical point, though, is that if Stupak's eventual vote for the PPACA was a sincere vote in the face of electoral pressure to do otherwise, then the party pressure applied to him and to his antiabortion colleagues was preference-preserving influence. Moreover, since nobody in the House cast a *pivotal* vote, that pressure was applied for the purpose of solving the collective action problem of sincere voting. The result was another increment by which legislators adopted somewhat more ideologically noncentrist, electorally suboptimal locations in the policy space. In exchange for doing so, legislators achieved a critical policy objective that could not have been achieved by an electorally diverse caucus with noncentrist policy preferences in the *absence* of party discipline.

That is exactly what the model in chapter 4 predicts. Moreover, it is difficult to imagine what threats would have been necessary to get Stupak and his colleagues to cast career-ending, nonpivotal votes that might have been insincere. Stupak and his associates voted yes because Democratic leadership solved a collective action problem of sincere voting and prevented Stupak and his colleagues from "free-riding" by casting electorally optimal, insincere votes in the hope that there would be too few other free-riders to cause the bill to fail.

Unfortunately, there is a critical question we cannot answer. What did Obama, Pelosi, or anyone else say to Stupak to get him *not* to free-ride? Negotiations occurred behind closed doors, for obvious necessity, which is the empirical challenge. The model in chapter 4 focused on how parties can use the threat of modest punishments to induce sincere votes, but there are actually several ways that the Democratic Party might have solved this particular collective action problem: threats, rewards, suasion, or coordination.

Could Pelosi have threatened a severe punishment if Stupak had voted against the PPACA? Possibly, but since Stupak decided not to even *try* to hold onto his seat in the 2010 election, it is difficult to imagine what such a threat may have been. After all, what would have been worse than losing his seat entirely?

A tangible reward seems equally unlikely, given both Stupak's subsequent retirement and the fact that we have not observed any specific reward in the aftermath. Most former members of Congress have little difficulty procuring lucrative lobbying jobs on their own anyway.

Moral suasion is another possibility. Party leaders may simply have convinced Stupak and his associates that they could not bring *themselves* to defeat the PPACA through free-riding, and they may have done so without any tangible offering. The question, then, is whether or not the vote was ultimately any different from the 1993 Omnibus Budget Reconciliation Act. But it was, for the very reason that the final tally was 219 to 212. Even if Stupak and his 11 supposed associates wanted to pass the bill, *voting* for it remained a coordination problem. Some, but not all, could vote no, and nobody wanted to be the one who cast the *pivotal* vote the way that Margolies-Mezvinsky did in 1993.

So there was a preference for an overall tally in which the bill passed by *more* of a margin than 216 to 215. That way, nobody was in the position of Margolies-Mezvinsky, taking the full brunt of the *blame*, which is the corollary to being pivotal. The problem, then, was that *some* could be allowed to free-ride by voting no, and determining who got to vote no was a coordination problem.

Whatever *else* happened when Stupak negotiated with Obama and Pelosi, that coordination problem was resolved in the process, and that is what matters most in terms of the basic logic of party government and spatial positioning. Margolies-Mezvinsky had no need to coordinate with anyone because she was pivotal. Stupak and his colleagues, though, at the very least needed a coordination mechanism, and whatever else party leadership did, it served that purpose. Sincere voting on the PPACA for Stupak and his colleagues was a collective action problem, solved by party leaders applying preference-preserving influence. The result was that legislators incrementally moved toward more ideologically extreme and electorally suboptimal positions in the policy space. Some lost their seats as a result. However, they got a policy outcome that they wouldn't have gotten without parties to solve the collective action problem of sincere voting. The Patient Protection and Affordable Care Act of 2010 passed the House of Representatives because the Democratic Party leadership solved a collective action problem faced by a caucus with similar personal preferences on the outcome of the roll call vote, but with diverse electoral pressures. The result was a policy victory that came at the cost of legislators adopting incrementally more extreme and electorally suboptimal positions in the policy space. However, the caucus was willing to accept this in order to achieve a policy victory that it could not have achieved otherwise.

One could argue that the outcome could have been achieved without party influence if individual legislators simply didn't know whether or not they would be pivotal. Thus, they would have to weigh probabilistic policy benefits against electoral costs. After all, prior to the final floor vote, many Democrats didn't know whether or not they had a majority. Thus, Stupak (and his colleagues) may have voted sincerely, not because of party pressure, but because the cost of a policy loss, multiplied by the probability of being pivotal, was greater than the electoral cost of voting for the PPACA. If so, no party influence would have been necessary.

The problem with this argument is that the roll call voting process is a sequential one. A roll call vote is not a one-stage, simultaneous move game. While the *order* of moves is indeterminate, legislators vote knowing how previous legislators have already voted. With 431 voting Representatives, the bill needed 216 votes. The final three yes votes were cast by legislators who knew that they were not casting the pivotal vote. They voted yes anyway. Additionally, any one *previous* yes vote could have been switched to an electorally optimal no vote once the number of yes votes reached *217*. Legislators voting yes in the end knew the electoral stakes. They knew they weren't pivotal. They voted yes anyway. For those representing centrist districts, that cannot have been for electoral reasons because it was clearly electorally dangerous. It cannot have been for policy reasons because nonpivotal legislators vote without policy consequences by definition. They were simply deciding not to free-ride, which only makes sense when parties solve the collective action problem of sincere voting, thereby creating incentives for legislators to incrementally adopt ideologically extreme, electorally suboptimal positions in the policy space.

Party Influence at Work

The theoretical model presented in chapter 4 showed how an ideologically homogeneous, but electorally diverse and policy-motivated, caucus has incentives to delegate power to party leadership to solve the collective action problem of sincere voting. It is only when parties act as preference-preserving institutions that such a problem can be solved, allowing the electorally diverse caucus to achieve its noncentrist policy goals. That setup closely resembles both the context and resolution of the House vote on healthcare reform in 2010. While the House Democratic caucus in 2010 was relatively unified in its desire to pass a healthcare reform bill, many members represented districts in which voting yes would be electorally dangerous. That created a collective action problem for wavering Democrats like Bart Stupak and his antiabortion colleagues. Without the Democratic Party to solve that collective action problem, through coordination if nothing else, healthcare reform could not have passed. The effect, though, was that Democratic legislators adopted incrementally more extreme, electorally dangerous positions in the policy space. They simply valued policy outcomes enough to outweigh those electoral costs. Nevertheless, those policy preferences would have been irrelevant without party influence because in the absence of a party to solve the collective action problem, any one wavering Democrat should have voted no for electoral reasons, regardless of the strength of their policy preferences, because if legislators aren't pivotal, their policy preferences shouldn't matter. That is exactly why policy-motivated legislators empower parties.

This puts legislative parties in a strange light. Their influence exists, at least in part, to counteract electoral pressure. In the process, then, political parties arguably distort representation, and they do so at the behest of their own members.

However, all of the analysis so far has been based on Congress before the 2010 election, and the divided government period from 2011 through 2016 presented a somewhat different picture of the relationship between Republican Party leadership, elections, and spatial positioning. Chapter 7 will examine what changed during that period, why, and what that means for the incremental positioning model.

7

Extreme Reversion Points and Party Leadership from 2011 through 2016

The unified spatial model derived in chapter 4 suggests that party leaders pressure legislators to adhere to their sincere, noncentrist policy preferences in the face of electoral pressure to vote as centrists, and while chapters 5 and 6 show this dynamic at work in the pre-2011 era, a great deal changed in Congress after the 2010 election. After "Tea Party" activists helped the Republicans regain the House majority in the 2010 wave election, members of the Republican caucus seemed to take little interest in positioning themselves as centrists, and hence had no need for party pressure to counteract the centripetal forces of general elections. Many seemed more concerned that a lack of conservative purity would result in a successful primary challenge. Chapter 5 argued that such fears would be overreactions, but that doesn't mean the fears aren't real and powerful motivators, even if they represent a break from older patterns.

Faced with a primary-fearing caucus, Republican leadership in the House has sometimes looked more like it is striving to temper extremism rather than to counteract electoral pressure toward centrism. Most visibly, Speaker John Boehner (R-OH) allowed, and perhaps even encouraged, minorities within his caucus to join coalitions with Democrats on pieces of "must pass" legislation when those members may have wanted to vote against these bipartisan agreements in order to establish their conservative bona fides so as to avoid primary challenges. This culminated in Speaker Boehner stepping down, pressured by the House Freedom Caucus, in retaliation for his approach to leadership. That implies a different dynamic between leadership and rank-and-file than the model presented in chapter 4. Yet, as we shall see, what happened is easily explainable within the context of the general model once we understand how the 2010 election invalidated the agenda assumptions of the model in chapter 4.

As we shall see, the dynamic that developed between Republican Party leadership and its caucus in the House after the 2010 election resulted, not merely from a newfound fear of primaries, but from a shift in the substantive legislative agenda in response to the combination of extreme polarization and divided control of government. Divided government is nothing new. Throughout much of the post–World War II period, it has even been the norm. As Mayhew (1991) pointed out, that hasn't historically been an obstacle to legislative action, but the high level of legislative activity during the Nixon and Ford administrations does not mean that divided government is irrelevant to policymaking. Both Nixon and Ford were liberals by the standards of the modern Republican Party, and the Democratic majorities they faced in Congress included many moderate-to-conservative southern Democrats, who are now gone. When President Nixon was inaugurated, Representative John McCormack (D-MA) held the Speaker's gavel. The two actually served in the House together, as late as the 81st Congress.[1] During that Congress, Representative McCormack's DW-NOMINATE score was −.2540. Representative Nixon's score was .1540. In contrast, consider Representative Boehner's score before becoming Speaker. In the 111th Congress, his DW-NOMINATE score was .7260. In his years before becoming President, Senator Obama's 110th Senate DW-NOMINATE score was −.3910. Cross-chamber comparisons notwithstanding, the chasm between President Obama and Speaker Boehner dwarfed the modest gap between President Nixon and Speaker McCormack.

Just because significant legislation could pass during Nixon's presidency, then, does not mean that we should have expected significant cooperation between President Obama and Speaker Boehner. Rather, the gap between Obama and Boehner's caucus was so large that it is difficult to imagine the kind of normal legislation that might unite them. We should not be surprised, then, by the absence of any major legislative activity on "normal" bills from 2011 through 2016. There would have been no point. Any bill moving policy to the left wouldn't get a vote in Congress, and any bill moving policy to the right would have been vetoed. Moreover, while the second dimension of NOMINATE was rather important in 1969, too much has collapsed onto the first dimension for there to be significant room for compromise. Divided government alone doesn't stop normal legislative activity. Introducing divided government into an era of extreme polarization, though, virtually shuts down the legislative process.

Rather than working on substantive legislation on high-profile issues like immigration reform, then, the House primarily started holding a series of symbolic votes to "repeal Obamacare." With no chance of overriding a presidential veto, nor even passing the Senate, though, these votes were no more than what Mayhew (1974) would call "position-taking." The combination of extreme polarization and divided government, introduced in 2011 at a level previously unseen, simply shut down the legislative process.

But it couldn't remove the need for legislation. Congress still needed to approve spending and raise the "debt ceiling." However, reduced to these kinds of votes, the legislative agenda winds up violating the agenda assumptions from chapter 4. These votes do not involve conventional "status quo" points. Rather, they involve ideologically extreme "reversion" points, and while normal legislative votes have cut points located between the opposing leaders' ideal points, bills dealing with ideologically extreme reversion points can have cut points outside that region, which changes the legislative dynamic. Consequently, the majority caucus was fractured in unusual ways by these agenda items, causing a shift in the direction of party pressure, but not in the fundamental purpose of party leadership, which is to solve a collective action problem of sincere voting for a caucus facing electoral pressure to vote insincerely. Nevertheless, by repeatedly fracturing the caucus, Speaker Boehner necessarily alienated the segment pressured into voting with the Democrats because of the increased (perceived) risk of primary challenges, while also alienating the faction voting against the deals for passing policies they disliked. That left Boehner's speakership intrinsically unstable, making his 2015 resignation an inevitable consequence.

Status Quo Points versus Reversion Points and a Primary-Fearing Caucus

While the tension between Speaker Boehner and his caucus throughout his five-year term broke from the pattern described in the unified model, let us consider two adaptations to the underlying model to see how they would change the dynamic between leadership and rank-and-file. First, consider the innocuous Assumption 5 from chapter 4. The assumption stated that there would be no roll call votes with cut points located outside the range between the opposing leaders' ideal points. The reasoning was quite simple. If a status quo point were located outside that range, it should be unstable because there would be broad, bipartisan support to change it, and if there were an alternative point outside that region, the bill would be modified with amendments before reaching the floor. Let us focus, though, on the first part of that reasoning. It is not that a status quo point *cannot* exist outside the range between the leaders' ideal points. They simply shouldn't be the norm because of their instability. However, what if, instead of bills that pit an alternative against a status quo point, we observe a bill that pits an alternative against a *reversion* point—some location other than the status quo to which policy will revert if no bill passes. Some legislation has a sunset provision, such that failure to renew the legislation causes policy to revert, not to the status quo, but to the prebill state of policy. In fact, all nonentitlement spending[2] essentially has such a sunset provision. Nonentitlement spending is appropriated from fiscal year to fiscal year, and without new spending approved, the program or agency runs out of money. If no appropriations are made, then, the government gradually shuts down due to lack of funding. In the context of appropriations, then, failure to approve new spending results, not in the maintenance of current spending levels, but in a reversion to a shutdown. A set of appropriations pit the spending levels in the bill against a shutdown rather than against the status quo.

Even more dangerously, consider the "debt ceiling." Statutory law instructs the Treasury to disburse money according to a combination of annual appropriations and existing entitlement programs whose funding is governed by preset formulae. However, Congress does not give the IRS the authority to collect enough money in tax revenue to pay for all of it, so the Treasury can only make all required disbursements if it issues bonds to raise money. However, a statutory law known as the "debt ceiling" caps the value of outstanding bonds that the Treasury may issue. If that statutory limit is reached, the Treasury must either violate Congress's instructions on disbursements or violate Congress's instructions on the debt ceiling by issuing more bonds in order to bring in the money that it has been instructed to disburse. Moreover, failing to disburse that money could have dire economic consequences. If bondholders were not paid, investors would lose confidence in the security of Treasury bonds, causing interest rates to spike. Even if bondholders were paid, though, many people owed money by the government would not be, which would both directly reduce economic activity and potentially cause bondholders to lose confidence anyway, since they would worry that someday they may not be paid. Either way, the result would be a shrinking economy, a flood of lawsuits, increased interest rates, and the indirect economic consequences of rising interest rates.

To most political actors, government shutdowns and debt ceiling breaches are unacceptable calamities. Not so for extreme conservatives. To extreme conservatives, a shutdown ranges from acceptable to a goal in and of itself, and many express skepticism that a debt ceiling breach would cause any real harm. Consider, for example, Representative Ted Yoho (R-FL), whose DW-NOMINATE score in the 113th Congress was .9560. He claimed in 2013 that a debt ceiling breach would "bring stability to world markets."

What, then, is the best way to characterize a shutdown or debt ceiling breach within a spatial model? We could construct a simple model of brinksmanship in which everyone loses some utility if a deal is not reached in time, but that presumes that all actors expect to lose utility from the absence of a deal. Those like Representative Yoho clearly don't expect to lose from failure to raise the debt ceiling, and those actors are on one extreme end of the ideological spectrum. Thus, we should not assume that a shutdown or a debt ceiling breach imposes costs on all actors. Rather, we should treat these outcomes as ideologically extreme *reversion* points, appealing to extreme conservatives like Representative Yoho, but abhorrent to liberals, and generally anyone who isn't as extreme as

Yoho. In fact, then, one must treat these outcomes as reversion points that are even more conservative than Republican leadership in the current era. Thus, interactions to prevent shutdowns and debt ceiling breaches violate Assumption 5 from chapter 4.

Next, consider that all of the analysis in chapters 2 through 6 focused on centrist electoral pressure. While chapter 5 discussed how the current fear among Republican officials of "getting primaried" is an unreasonable overreaction to a few odd instances, that doesn't mean the fear isn't real.

In particular, we must apply fear of primaries to votes on government shutdowns and debt ceiling increases. The Tea Party activists who helped Republicans retake the House majority in 2010 styled themselves as antigovernment activists, and if so, then they must applaud the concept of a government shutdown. Similarly, Republicans appealed to these activists on the grounds that government spending must be reined in, and the phrase "Raise the debt ceiling" is not one likely to appeal to such voters. Thus, it is on shutdowns and debt ceiling increases that these voters might be most activated.

Thus, if we want to understand the relationship between Republican leaders and their rank-and-file in the divided government period from 2011 through 2016, we must make two changes to the basic model. We must replace centrist status quo points with ideologically extreme *reversion* points, and we must replace centrist electoral pressure with extremist electoral pressure, at least within the Republican caucus.

Revising the Model in the Context of Agenda Changes

In order to see how these changes would affect the dynamic between party leadership and its rank-and-file, let us consider the base case. Chapter 4 examined the legislative process when the majority party attempts to move policy toward the preferences of its ideologically homogeneous caucus, and away from the center. Figure 7.1 represents such an agenda item.

Figure 7.1 shows the relationship between five locations along a policy dimension: the status quo point (SQ) reflecting the current state of policy, the alternative point (A) reflecting where policy would be if the bill in question passed, the location of the floor median (FM), the cut point, which is the midpoint between the status quo and alternative points, and a simplification labeled the "electoral divide." The "electoral divide" represents a systemic gap between legislators' personal preferences and their constituents' preferences, creating the collective action problem of sincere voting. Let us assume that legislators' ideal points are correlated with their district median voters' ideal points, but that they

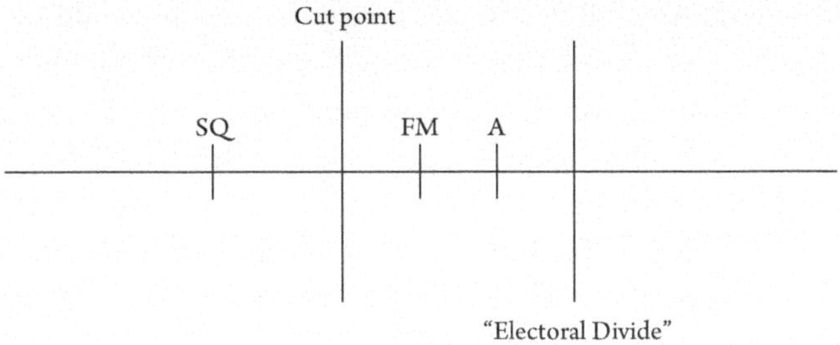

Figure 7.1 The collective action problem of sincere voting

are systematically more extreme than their district median voters. We can formalize this bias as follows. Any legislator with an ideal point to the right of the electoral divide represents a district in which the median voter is sufficiently extreme to prefer the alternative to the status quo. However, any legislator with an ideal point to the left of the electoral divide represents a district in which the median voter prefers the status quo. Thus, legislators whose ideal points are located between the cut point and the electoral divide are cross-pressured. They personally prefer the alternative to the status quo, but voting for the alternative would incur an electoral penalty because they represent more moderate districts in which the voters prefer the status quo. Thus, voting for the alternative would be electorally suboptimal because it would be an incremental step toward adopting a location in the policy space that is too ideologically extreme. This is just a graphical representation of the basic model from chapter 4.

Since the floor median's location is between the cut point and the electoral divide, she and similarly positioned legislators are subject to a collective action problem of sincere voting (the collective action problem exists as long as there is a *group* of similarly situated legislators in that range). According to the model in chapter 4, if the caucus is policy motivated, party leaders solve that collective action problem by applying preference-preserving influence to legislators whose ideal points fall between the cut point and the electoral divide.

Chapters 5 and 6 argued that this dynamic explains the process by which members of Congress from centrist districts gradually moved away from their district median voters in the post–World War II period, and the method by which the House of Representatives passed the Patient Protection and Affordable Care Act in 2010. However, what would we expect to happen when we combine extreme polarization, divided government, and a House majority that fears primaries more than general elections, that is, the period from 2011 through 2016?

First, consider the agenda. Under such conditions, the House has no incentive to take up any regular legislation, moving policy from a status quo point to an alternative point. Any bill that moves policy to the right would be vetoed by President Obama, and any bill that moves policy to the left wouldn't make it to the floor of the House. However, Congress could not stop legislating altogether. New spending would still need to be approved, and the debt ceiling would still need to be raised. Under pre-2010 circumstances, these were routine pieces of legislation, but when purged of any other substantive items, the legislative agenda necessarily centers on these types of bills, with policy changes only possible when those changes are attached to must-pass legislation like debt ceiling increases. A legislative agenda built around items such as healthcare reform is very different from a legislative agenda built entirely around averting a debt ceiling breach, particularly for a primary-fearing caucus with a large contingent of extremists, like the modern Republican caucus. So we can modify Figure 7.1 to reflect a vote involving an ideologically extreme *reversion* point and electoral pressure toward extremism. Consider Figure 7.2.

In Figure 7.2, rather than a relatively centrist status quo point, there is a *reversion* point, for example, breaching the debt ceiling, that is so ideologically extreme that only the most extreme conservatives, like Representative Yoho, sincerely prefer it to the alternative point, which is still to the right of the floor median's ideal point. In this case, though, the electoral divide is to the left of the floor median's ideal point. Legislators whose ideal points lie to the left of the electoral divide are Democrats. Since they do not run in Republican primaries, they do not fear losing primaries to Tea Party challengers if they vote to raise the debt ceiling. While Democrats have looked for opportunities to vote against debt ceiling increases in the past, they have been willing to support increases at least since 2010.

The challenge in the 2011 through 2016 period was that there weren't enough of these legislators to pass a debt ceiling increase on their own. Legislators whose ideal points lie between the electoral divide and the cut point prefer the still-very-conservative alternative to the reversion point, but fear electoral punishment if they vote for the alternative because of the now rightward electoral

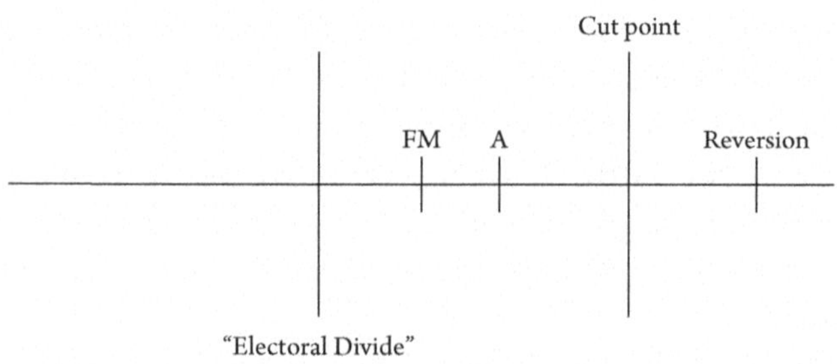

Figure 7.2 The collective action problem with an extreme reversion point

pressure of primaries, as they perceive it (wrongly or not). Moreover, since the floor median falls in that range, we have the familiar collective action problem of sincere voting as long as there is a cohort around the floor median with similar cross-pressures. If only legislators with ideal points to the left of the electoral divide are voting for the alternative, then any one cross-pressured legislator who sides with the minority will be casting a sincere but electorally suboptimal, nonpivotal roll call vote. That would mean paying an electoral penalty for no policy gain. If the reversion point is to be avoided, some of these cross-pressured legislators must join with the Democrats by casting sincere but electorally suboptimal roll call votes for the alternative, despite their fears of "getting primaried" if they do so. The only way for that to happen is if the majority party leadership applies preference-preserving influence to some subset of legislators located between the cut point and the electoral divide, thereby solving the collective action problem of sincere voting.

Ironically, though, depending on the location of the cut point, this might mean pushing a small number of majority party members to vote against the majority of their caucus. The structural factor that creates this odd dynamic is that the majority party is not trying to move policy away from the center. It is trying to avoid an extreme *reversion* point with which some of the caucus members are actually comfortable. Regardless, majority party leaders must address the circumstance by using preference-preserving influence to solve a collective action problem of sincere voting, just as in chapter 4.

There are two difficulties for a party leader handling a circumstance like the one shown in Figure 7.2. The first, and most obvious, is that avoiding the reversion point actually alienates a subset of the caucus on policy grounds since caucus members who have ideal points to the right of the cut point actually prefer the reversion point to the alternative point. The larger that faction is, the less tolerant the caucus will be of party influence, and the greater the risk of backlash against the Speaker who wields that influence. Second, the precise circumstances that create the division in the majority caucus also limit the Speaker's disciplinary options. In general, manipulation of the committee assignment process can be a critical tool, and it even allows for variation in the severity of punishment. Removing legislators from powerful committees, for example, is a strong measure, but refusing to grant a transfer to a modestly better committee is a weaker measure. However, committee assignments are primarily important when the House is taking up regular legislation. Since floor amendments are generally limited in the House by rules constructed by the Rules Committee, a legislator who wants a significant degree of control over the content of a piece of legislation needs to be on a committee with jurisdiction over the bill. However, if this circumstance occurs because the House has stopped working on normal legislation because the gap between the president and the House majority is too large to bother, then the same circumstance that creates the problem limits the value of the incentives party leaders have to offer. If even spending is handled with a continuing resolution negotiated on the floor at the last minute, then even a seat on the Appropriations Committee is of diminished value. Thus, when faced with a circumstance like Figure 7.2, the Speaker has a legislative agenda built around ideologically extreme reversion points, and faces the challenge of leading by *fracturing* his caucus when even his strongest disciplinary tools aren't as strong because of the exact same circumstances that create the problem in the first place.

Moreover, party leadership in a circumstance like Figure 7.2 must be inherently unstable.[3] In conventional versions of CPG, the critical condition for party strength is ideological homogeneity. The model in chapter 4 maintains ideological homogeneity as a condition, but to that adds electoral heterogeneity. However, Figure 7.2 represents the opposite circumstance—ideological heterogeneity and electoral homogeneity. Universal fear of primary challenges, to the degree that it exists, creates a uniformity of electoral pressure regardless of how dissimilar caucus members' general election constituencies may be, while the introduction of an ideologically extreme reversion point splits the caucus on policy grounds. Since the function of party leadership here is to exacerbate that policy division, the Speaker necessarily makes enemies in the process.

Note, though, that in many ways the circumstance is similar to the model in chapter 4. The job of the Speaker remains to solve the collective action problem of sincere voting created by members having electoral incentives to vote against their own policy preferences. However, instead of party leadership enforcing unity in order to achieve a policy victory, party leadership exploits, and indeed creates, *dis*unity. The difference follows from the diminished legislative agenda resulting from the combination of extreme polarization and divided government, and a fear of primaries rather than general elections, both of which came into play after the 2010 election. On theoretical grounds, though, following from the logic of the unified model, we should expect the politics of extreme reversion points to pit party leaders against their own most extreme elements rather than against centrist electoral pressure. As we shall see next, this is precisely what happened during the divided government period from 2011 through 2016.

The Debt Ceiling and the House of Representatives from 2011 through 2016

As we have seen, replacing centrist status quo points with ideologically extreme reversion points, and replacing centrist electoral pressure with extremist pressure from primaries, creates a very different dynamic between leaders and their rank-and-file. The task of the leader becomes fracturing the party rather than unifying it to solve what remains a collective action problem of sincere voting. More importantly, the narrative of conflict within the House Republican caucus during the period of divided government under President Obama closely resembles the simplified model in Figure 7.2. Consider what happened.

With a weak economy, a relatively unpopular Democratic president, and a controversial healthcare reform law passed earlier that year, the midterm elections of 2010 switched control of the House of Representatives to the Republican Party. However, the Republicans failed to acquire a majority in the Senate, much less veto-proof majorities in both chambers. Consequently, their midterm seat gains did not give Republicans the ability to enact substantive policy changes. While the House engaged in frequent position-taking votes to "repeal Obamacare" that Republicans knew wouldn't lead anywhere, along with a series of oversight hearings, most visibly on the 2012 attack on the US Embassy in Benghazi, Libya, the House did not put any real effort into immigration reform, tax reform, or other complicated policy issues requiring negotiation with the Democratic administration. The ideological chasm between the administration and the House majority would have made such efforts futile.

From the perspective of actual policy outcomes, the important votes from 2011 through 2016 were on issues such as the continued funding of government and the debt ceiling. It was on these votes that critical internal divisions within the Republican caucus developed in ways that closely resembled Figure 7.2, extrapolated from the unified model.

The debt ceiling was the most difficult issue for the Republican Party after 2010. Raising the debt ceiling has historically been politically uncomfortable, but this has been especially true for Republicans since the 2008 financial crisis. The phrase "raise the debt ceiling" sounds analogous to a debt-ridden family asking its credit card company to raise its credit limit so that members of the household can continue recklessly spending themselves deeper into debt. Voting against raising the debt ceiling, then, is a convenient position-taking method to define oneself as "fiscally responsible." The flaw in the analogy, of course, is that once a family "maxes out" its credit cards, merchants will reject those cards, thereby preventing the family from incurring new financial obligations beyond the accruing interest and late penalties. In contrast, a debt ceiling breach only occurs if Congress chooses to incur financial obligations that it doesn't give the Treasury the authority to cover. Those financial obligations remain legally binding even if the Treasury has reached its borrowing authority, which is precisely what makes a debt ceiling breach so dangerous. Nevertheless, minority party legislators have regularly looked for opportunities to vote against a debt ceiling increase in order to take a symbolic position in favor of balanced budgets, thereby forcing majority party members to cast the uncomfortable yes votes. As a Senator, Barack Obama was among the many to do so.

For the new Republican majority in the 112th House, though, the problem was more complicated. In the aftermath of the 2008 financial crisis, Republican politicians rallied around the claim that excessive government spending was a primary cause of the crisis and that resolving the crisis necessitated reducing government expenditures. Moreover, the Tea Party activists who formed the electoral base of the Republican Party frequently insisted that their primary political objective was to reduce government spending. Thus, Republican legislators feared that if they voted to raise the debt ceiling, their electoral base would see it as a betrayal. In principle, that could lead to a Tea Party primary challenge.

Despite that, Republican Party leadership, extreme though it was (Speaker Boehner's DW-NOMINATE score prior to taking the gavel was .7260), publicly stated that the debt ceiling must be raised, in agreement with the entire Democratic Party. Then-Senate Minority Leader Mitch McConnell (R-KY) even proposed a change to the debt ceiling process. He suggested that power be given to the president to raise the debt ceiling unilaterally, with Congress retaining the power to block the increase through a piece of legislation that would need to overcome a subsequent presidential veto in order to prevent the debt ceiling increase. The point of McConnell's proposal was to ensure that the debt ceiling would be raised, while allowing Republicans to engage in another position-taking exercise by voting to overrule it, knowing that they could not override the inevitable veto. Clearly, Republican Party leadership did not want to breach the debt ceiling.

In contrast, recall Representative Yoho's assertion that it would be beneficial to breach the debt ceiling. Representative Yoho's DW-NOMINATE score in the 113th Congress was .9560. If we accept Representative Yoho's statement as a sincere reflection of his personal beliefs (and in his case, there is no reason *not* to do so), the implication is that the most extreme conservatives, like the activist base, opposed raising the debt ceiling because their ideal points were close to the reversion point of a debt ceiling breach. Everyone else's ideal points were too far from that reversion point for such an outcome to be acceptable.

The problem for the majority caucus was that members all felt primary pressure to vote against raising the debt ceiling. But, as the majority in the House, they needed to supply at least some Republican votes for a debt ceiling increase. Otherwise, they would have to face the economic consequences of a debt ceiling breach and the political consequences of that economic damage.

In the summer of 2011, Treasury warned that a debt ceiling breach was imminent, and the new Republican majority needed to find a way to avert a breach that would minimize the risk of a backlash from its electoral base. Republican Party leaders managed to thread that needle by using the debt ceiling as what Mitch McConnell called "a hostage that's worth ransoming." Rather than face the risk of a Tea Party backlash for simply raising the debt ceiling, Republican leaders insisted that they would refuse to raise the debt ceiling unless President Obama and Senate Democrats agreed to a set of Republican demands. Specifically, Speaker Boehner demanded that every dollar increase in the debt ceiling be accompanied by a dollar in spending cuts.[4] If Boehner could convince Democrats to make concessions on spending, then Republican leaders could pacify Tea Party opposition to raising the debt ceiling with the concessions acquired by doing so. Moreover, Republican leaders could claim that they didn't actually *want* to raise the debt ceiling—they only agreed to do so in order to achieve other goals.

Extracting concessions, though, requires being willing to vote for a deal that includes raising the debt ceiling once the opposing party agrees to concessions. However, as deliberations began in the 112th Congress, many Republicans in the House issued a blanket refusal to vote for any debt ceiling increase. Some simply feared that the Democratic concessions would not satisfy Tea Party activists, and others likely agreed with those activists. This, too, fits with the model in Figure 7.2. Republicans with ideal points to the right of the cut point would be implacable, and those to the right of the electoral divide would be at least difficult to corral.

To make matters more complicated, many Democrats were unwilling to agree to the types of concessions Republicans were demanding for a debt ceiling increase. The strategic challenge that Speaker Boehner first faced in the 112th Congress, then, was to cobble together a bipartisan coalition in the midst of a historically polarized Congress, unless he wanted Republicans to take the blame for the fallout from not raising the debt ceiling. That required convincing Republican legislators to vote for a bill containing a debt ceiling increase, even if doing so might risk a primary challenge fueled by Tea Party activists. The challenge, then, was to attract enough votes from legislators with ideal points between the electoral divide and the cut point to supplement the Democratic votes. The problem was that if no one was pivotal, then no one had any policy incentive to vote for the deal. Everyone wanted to free-ride by voting against the bipartisan deal in order to stave off a primary challenge. Without party leadership to solve that collective action problem, the deal would fail, the debt ceiling would be breached, and the entire party would suffer the consequences. This should sound familiar because, in essence, it is the same story as the unified model in chapter 4 and the challenge faced by Democrats trying to pass the Patient Protection and Affordable Care Act, but with a reversal of the direction of pressure.

In August 2011, the new Republican majority passed its first debt ceiling increase as part of the Budget Control Act of 2011. The bill matched a debt ceiling increase with a set of deficit reduction measures equal to the value of the debt ceiling increase. The bill specified some spending cuts directly and mandated the creation of a "supercommittee" to propose the rest of the deficit reduction measures with the fallback position that if the supercommittee failed, a set of broad and intentionally reckless spending cuts known as "sequestration" would go into effect.[5] The Budget Control Act passed the House by a vote of 269-161, with both caucuses divided. Democrats were divided because many balked at the magnitude of the spending cuts. More interesting, though, were the divisions among House Republicans. Speaker Boehner considered the bill a victory, claiming to have achieved "98%" of what he wanted because the bill included extensive spending cuts and no tax increases. However, achieving that victory required convincing House Republicans to support the bill even though many of them either wanted even more extensive cuts, opposed raising the debt ceiling in principle, or both. Moreover, the fear of a Tea Party primary challenge, either immediately or at some indeterminate future point, made a simple no vote electorally appealing. Voting no can send a more ambiguous signal than voting yes because legislators can offer a variety of indistinct reasons for rejecting a bill, as opposed to the simple interpretation of a yes vote as a full-stop endorsement of the bill (Buchler 2017). Any Republican worried about either current or future ideological purity tests could simply vote no and claim that any deal was insufficiently conservative without stating what would have been required for a yes vote.

It was not clear until the end that Boehner would find enough Republican votes to pass the bill. In the end, with 95 Democrats voting yes and 95 voting no, Boehner needed 121 primary-fearing Republicans to support the deal (assuming 430 votes cast). He got 174. The most conservative faction of House Republicans voted against the deal, either because of the debt ceiling increase itself or because the spending cuts were insufficiently large, but Boehner found enough to pass the bill. Boehner solved the collective action problem, but he did so by dividing the caucus rather than unifying it, as in Figure 7.2.

While the bill did pass, there was so much uncertainty about its success leading up to the final vote that the ratings agency of Standard & Poor's reduced the credit rating of US Treasury bonds based on the fear that someday, the increasingly dysfunctional Congress might miss the deadline and permit a debt ceiling breach. Fearing both the economic and electoral consequences of that, it became important for Speaker Boehner to pose a credible threat to punish intransigent legislators who refused to vote for deals to increase the debt ceiling, ensuring his ability to solve collective action problems.

Shortly after the 2012 elections, House Republicans shuffled committee assignments, and three very conservative legislators who had bucked party leadership on important votes such as the debt ceiling were stripped of their important committee assignments—Justin Amash from Michigan, Tim Huelskamp from Kansas,[6] and Dave Schweikert from Arizona.[7] The event sticks out for two reasons—the specific targets and the severity of the punishment. When Democrats removed three committee chairs after relaxing seniority rules as part of the 1970s reforms, they were directing pressure at the moderate-to-conservative members of the caucus, which is how one normally thinks of party discipline. However, Amash, Huelskamp, and Schweikert were extremists, not centrists. Also, the model in chapter 4 was based on the application of very modest disciplinary threats to achieve significant policy effects, but these were severe punishments.

However, since it was with the conservative flank that Boehner found himself in conflict, it was among that flank that he needed to make examples, and since normal disciplinary tools had diminished value, only harsh punishments would do. Of course, given their positions on the ideological spectrum, Amash, Huelskamp, and Schweikert were likely opponents on any debt ceiling deal. If they were already adversaries, little would be lost by punishing them. The point, though, was to demonstrate willingness to carry out the harsh punishments necessary to achieve compliance on the difficult issue of the debt ceiling. The real targets, then, were not necessarily the *far*-right flank, but the legislators between the electoral divide and the cut point. Boehner could not alienate *them* by punishing any of them at the start of the 113th Congress because he might need their votes, but he needed to demonstrate his willingness to carry out such punishments to legislators in that range, and the least harm would be done by punishing those who were already alienated by the Budget Control Act in 2011.

Immediately afterward, the debt ceiling returned to the legislative agenda, with a breach looming in February 2013 unless Congress acted. Unlike in 2011, though, President Obama insisted that Republicans agree to a debt ceiling increase with no concessions. If they didn't, he would hold them publicly responsible for the economic consequences. Still, Republicans were nervous that if they voted to raise the debt ceiling, particularly without associated spending cuts, they would face primary challenges.

Despite that nervousness, having recently made examples of Amash, Huelskamp, and Schweikert, Speaker Boehner was able to secure Republican votes for a debt ceiling increase without spending cuts. Instead, the deal included a pair of symbolic concessions intended to allow Republicans to vote for an increase while saving face. First, according to the technical language of the bill, it did not "raise" the debt ceiling—the bill just "suspended" it. However, that was merely a rhetorical gimmick. The slightly less symbolic concession addressed a common Republican criticism of congressional Democrats and, in particular, Senate Democrats. Republicans objected to the fact that the Senate had not "passed a budget" in years, so the debt ceiling deal required Senate Democrats to "pass a budget" or else not receive their pay. However, the demand was merely that Senate Democrats pass a nonbinding resolution called "a budget," even though the elements of that nonbinding resolution were included, in binding form, in the Budget Control Act of 2011. Nevertheless, including the requirement in the debt ceiling deal allowed Republicans to claim to have won something significant without the Democrats actually giving up anything substantive. So would primary-fearing Republicans vote for a deal that included a face-saving symbolic concession but no spending cuts? Boehner found enough Republican votes—electorally *sub*optimal and *non*pivotal votes—to pass it by a vote of 285-144, with Republicans supporting the bill by a margin of 199-33. It should be noted that this was *more* Republican support than Boehner got for the 2011 Budget Control Act, even though that prior bill contained extensive policy concessions to Republicans, and the new bill contained none of any substance. Making examples of noncooperators had worked. Still, though, Boehner found himself at odds, not with the moderates in his caucus, but with the extremists, just as we would expect when a policy debate pits an alternative against an extreme reversion point rather than a centrist status quo point, as in Figure 7.2.

In October 2013, the debate played out somewhat differently. Failure to pass a continuing resolution by September 30, 2013, led to a partial government shutdown that began on October 1, and without a debt ceiling increase, the Treasury was predicted to run out of "extraordinary measures" to avoid a default on October 17. Consequently, negotiations to end the shutdown and raise the debt ceiling were combined into one process. After a debt ceiling increase with no real concessions earlier that year, Senator Ted Cruz (R-TX) convinced many of his Republican colleagues in the House, and a few in the Senate, to demand either a delay in implementation or defunding of the PPACA as a condition for reopening the government and raising the debt ceiling, but Democrats insisted that they would not offer any concessions in exchange for either reopening the government or raising the debt ceiling. As far as they were concerned, agreeing to a continuing resolution with spending levels set by the "sequester" was already enough of a concession, and acceding to more demands would simply reinforce the Republican view that the debt ceiling was, in Senator McConnell's words, "a hostage that's worth ransoming." Ultimately, the Republicans agreed to a continuing resolution and debt ceiling increase with no substantive concessions and hardly any face-saving symbolic concessions. Upon reaching an agreement, Speaker Boehner admitted that the Republicans had "lost" the battle, in stark contrast to how he had described the Budget Control Act in 2011. While the deal passed by 81-18 in the Senate, the House once again saw a divided Republican caucus. All present House Democrats voted for the bill (198), while a majority of House Republicans voted against it, by a margin of 144-87.

The split was a high-profile example of a "Hastert rule" violation. Party agenda control models in Congress, like the Cartel model, suggest that the majority party will prevent a bill from reaching a floor vote whenever two conditions are met: a majority within the majority caucus prefers the bill to fail, but enough majority party members would defect to allow the bill to pass should the bill receive a vote. In popular dialogue, this strategy has become known as the "Hastert rule," even though the practice of implementing it predates Speaker Dennis Hastert (R-IL). However, under Hastert's speakership, Republicans had a semiformalized agreement that any bill opposed by a majority of the majority caucus would not receive a floor vote. However, Speaker Boehner could not adhere to that "rule" because the only way to pass a debt ceiling increase in 2013 was to find most of the support from Democrats. Boehner repeatedly found himself in the position of having to pass some version of a bill in order to avert a policy disaster, even though his caucus was reluctant to vote for any deal. On an issue like the debt ceiling, even many conservatives sincerely wanted a bill to pass. They just feared the electoral consequences of voting for it because of a recently developed fear of Tea Party primary challenges. Part of the solution became abandoning the Hastert rule, allowing Democrats to cast many of the votes necessary for passage, allowing a significant number of Republicans to cast the electorally optimal no votes. Some Republicans needed to take the electoral risk, though. As is always the case when votes involve private negotiation, it is difficult to determine precisely how Speaker Boehner convinced enough Republicans to split off from their most conservative colleagues in order to cast what are likely sincere, electorally suboptimal, nonpivotal roll call votes. On some votes, particularly in October 2013, it was not clear that the Speaker needed to apply any pressure at all—he may simply have needed to allow a Hastert rule violation without provoking a revolt among the most conservative wing of his caucus. In any case, though, the challenge of "leading," not by keeping his caucus together, but by splitting it apart, may seem perplexing, but as Figure 7.2 demonstrates, it follows logically from bills that prevent ideologically extreme reversion points supported by a faction within the majority caucus. Even in this case, though, the role of leadership is to solve the collective action problem of sincere voting to allow at least a segment of the caucus to achieve a policy victory at an electoral cost. Substantively, that is only modestly different from the unified model in chapter 4 and follows logically from a violation of Assumption 5 in that model, resulting from the combination of extreme polarization and divided government and its effect on the legislative agenda, along with newfound primary fears.

On September 25, 2015, Speaker Boehner announced his intention to resign midsession. The announcement was met by ecstatic joy among the House Freedom Caucus, consisting of the most radical members of the Republican delegation, intent on opposing bipartisan deals. Since leading by fracturing the caucus and carrying out harsh punishments in a manner consistent with Figure 7.2 is an intrinsically unstable leadership model, no one should have been surprised.

The problem was that the job was a thankless one. Any subsequent Speaker would find himself in the same bind that Boehner did, needing to pass deals with most of the votes coming from Democrats in order to avoid shutdowns and debt ceiling breaches, which extreme conservatives found acceptable anyway. Doing so would alienate the radical wing of the party and ultimately put the new Speaker in the same position. Initially, Majority Leader Kevin McCarthy from California announced his intention to seek the speakership, and while the commentariat assumed he would win, a preliminary vote count demonstrated to Representative McCarthy that he did not have the necessary floor majority, due to opposition from the same radical wing that forced Boehner's resignation. Predictably, there was then a dearth of candidates. While some of the radicals in the party announced their interest, none had anywhere near the votes necessary, and nobody who had the votes wanted to run. Representative Paul Ryan from Wisconsin, then Chair of the Ways and Means Committee, emerged as the only candidate acceptable to enough caucus members to reach a floor majority. As architect of the so-called Ryan budget—a conservative wish list of tax cuts and loosely specified slashes to social spending—Ryan had enough of a reputation as a conservative to satisfy the Freedom Caucus, but as a tactical opponent of shutdowns and debt ceiling breaches, the institutionalists in the party trusted him not to drive the caucus off a policy cliff.

The problem was that Representative Ryan was smart enough to know that as Speaker he would have the same obligations as Boehner and suffer the same consequences. In order to entice Ryan to serve, then, Boehner needed to clear the table, so to speak, and pass a debt ceiling increase and budget outline to carry through to the 2016 election to prevent Ryan from being run out of office the same way Boehner was. The best-case scenario was that Republicans would win unified control in 2016, freeing the Speaker to act more normally (the consequences of the 2016 election will be addressed later in this chapter). The worst-case scenario was that it would be just a temporary fix, and the party could work it out later. The point, then, is that the deal under which Ryan assumed the speakership was predicated on recognizing the structural problems Boehner had faced and avoiding them, at least through the 2016 election.

All of this remains within the bounds of what we should expect given the general contours of the unified model, once we understand what happens when Assumption 5 is violated with a primary-fearing caucus. Because of the ideological gap between the House majority and the President during the divided government period from 2011 through 2016, there was no point taking up normal legislative activity. So Congress did little except avoid reversion points, such as debt ceiling breaches and (at least extended) government shutdowns. With the legislative agenda focusing on these items, Republican leaders were pitted against their own right flank because ideologically extreme *reversion* points divided the right flank from the rest of the chamber rather than Democrats from Republicans. Moreover, the shift in the electoral divide altered, rather than eliminated, the collective action problem of sincere voting. Rather than urging the center of the party to join the hardliners despite centrist general election pressure, party leadership needed to urge the center to break from the hardliners and join the minority party despite centrifugal primary pressure. Nevertheless, the central issue was a collective action problem of sincere voting that required stronger than normal disciplinary measures. The use of harsh disciplinary measures to fracture the party, though, necessarily made party leadership unstable, so Speaker Boehner was eventually forced to step down. To prevent Speaker Ryan from facing the same backlash, Speaker Boehner passed a debt ceiling increase carrying through 2016, hoping that the next election would eliminate divided government and hence eliminate the problem. All of this follows the dynamic suggested by Figure 7.2, which only required modest changes from the basic conceptual model presented in chapter 4.

A Simple Question of Agenda Control?

The previous discussion, built around Figure 7.2, describes a caucus defined by electoral homogeneity and ideological heterogeneity because of an agenda item with an ideologically extreme reversion point. In that circumstance, the party leader is tasked with solving a collective action problem of sincere voting among the relative moderates in the caucus. However, there is another model compatible with the GOP's intraparty divisions from 2011 through 2016. Consider Figure 7.3.

Now the electoral divide is actually to the right of the floor median. As long as the alternative is brought to the floor for a straight up-or-down vote, the bill will pass, and the ideologically extreme reversion point will be avoided. There is no collective action problem of sincere voting since a floor majority has both policy and electoral incentives to vote for the alternative. The challenge for the Speaker here is the question of whether or not to bring the bill to the floor. Doing so may be a violation of the Hastert rule. Any legislator to the right of the cut point prefers the reversion point to the alternative point, so any legislator to the right of the cut point will want the Speaker to block a floor vote. So if the Speaker *permits* a floor vote, she faces the possibility of a revolt for *not* exercising negative agenda control. Moreover, majority caucus members located between the electoral divide and the cut point, while sincerely preferring the alternative to the reversion point, may have electoral incentives to join in a revolt against the Speaker for her failure to *block* a floor vote. This, too, would be compatible with at least some post-2010 brinkmanship.

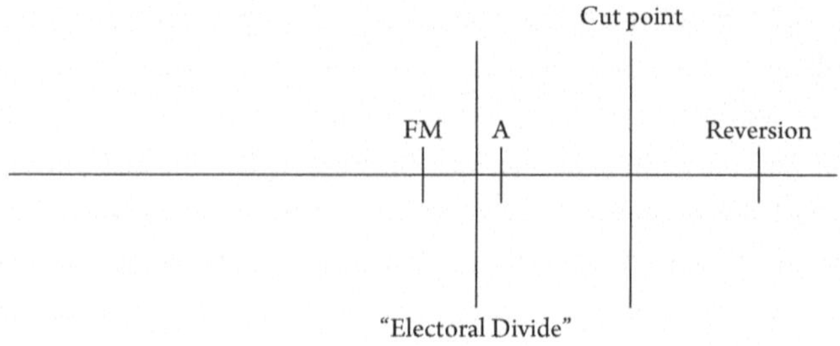

Figure 7.3 A roll call vote with an extreme reversion point but no collective action problem

The challenge is distinguishing between Figure 7.2 and Figure 7.3. The question is whether the constraint on the Speaker is that too many caucus members fear a primary backlash if they vote for a deal, as in Figure 7.2, or that failure to block a floor vote will alienate the extremists, as in Figure 7.3. The irony, of course, is that since the fear of a primary backlash is overstated, we cannot address the point by simply observing what happens to the moderates in the next primary. However, Figure 7.3 may be a better explanation for the October 2013 debt ceiling increase than the Budget Control Act. In October 2013, 87 Republicans joined 198 Democrats to vote for a debt ceiling increase without concessions, leaving 144 Republicans to vote no. With 429 voting legislators, only 215 votes were needed for passage. That means that of the 87 Republicans who voted for the deal, 70 could have voted against the deal without changing the result. If 80% of the Republican yes votes were unnecessary for passage, then had they wanted to vote no in order to fend off a primary challenge, many more could have done so. The fact that they didn't suggests that they weren't as worried about a Tea Party primary challenge, and the question was merely whether or not Speaker Boehner would permit a floor vote. Most likely, Republicans had been chastened by strongly negative polling for the Republican Party during the shutdown, which moved the electoral divide, or at least perceptions of its location. Thus, the challenge for Speaker Boehner in October 2013 was not the need to whip votes, but the need to violate standard agenda restrictions.

In 2011, though, Speaker Boehner had less room to maneuver. First, the conditions attached to the debt ceiling increase to bring Republican votes on board meant that Boehner could *not* count on a unified bloc of Democrats. That meant he needed far more Republican support. Moreover, the political environment in 2011 was very different from the environment in 2013. The 2011 Budget Control Act came to the floor at the height of the Tea Party, shortly after a record Republican landslide election, and without a government shutdown tilting the polls against Republicans nationwide. Thus, the collective action problem of sincere voting was a more significant factor in 2011 than in 2013. Where the

2011 vote broke from Figure 7.2, though, was that the Democrats were not unified in their support. As discussed earlier, though, this was because they didn't support the conditions attached to the debt ceiling increase, not because they didn't support the debt ceiling increase itself.[8] So Figure 7.2 represents how the *Republican* caucus viewed the 2011 debt ceiling increase rather than how the entire *House* viewed the bill.

In each case, though, the central point is that the tension between Republican rank-and-file and their leadership is created, not just by idiosyncratic primary pressure, but by a legislative agenda structured around ideologically extreme reversion points rather than centrist status quo points. Avoiding an ideologically extreme reversion point requires taking actions that alienate part of a caucus characterized by a form of ideological *hetero*geneity that is revealed by a legislative agenda stripped of "normal" legislation and focused on must-pass bills.

Ideological Homogeneity among House Republicans?

Since the 2010 election, the Republican Party has been in conflict with itself in the House of Representatives, and more broadly. Prior to 2010, general perception was that the Republican Party was the more unified, orderly party. Did the party change fundamentally in 2010 because of the Tea Party, as so many public commentators have suggested? The answer may be no. Recall one of the central insights from chapter 2—ideological distinctions are either hidden or revealed by the structure of the legislative agenda. Normal legislative agendas can obscure differences between party leaders and extremists within their own caucus because they involve relatively centrist status quo points. However, agenda items with extreme reversion points violate the agenda assumptions in chapter 4, thereby uncovering previously hidden ideological distinctions. The Republican caucus after the 2010 election may have been no more ideologically heterogeneous than it was before 2010. The distinctions were simply hidden by the structure of the agenda in the 111th House and revealed by the agenda in the 112th House. That revelation, though, created a collective action problem. More importantly for Boehner, solving that collective action problem alienated the more extreme caucus members, putting the process starkly at odds with standard models of CPG, or even Cartel politics. Since he led by fracturing the party rather than unifying it to solve a modified collective action problem of sincere voting, Speaker Boehner's position was inherently unstable, putting him at risk of a revolt rather different from the one that brought down Joe Cannon a century earlier. That revolt began at the start of the 113th Congress with symbolic votes against his continued speakership, and culminated in 2015 with his forced resignation, to the jubilation of the House Freedom Caucus. That is precisely what we would expect given Figure 7.2 and the basic principles of the updated CPG model. The point, though, is that while the Republican Party is not truly ideologically homogeneous, it may have been just as divided *before* the 2010 election and simply had those divisions hidden by normal legislative agendas. Whether those divisions can be tamped down again is another question, to be addressed later in this chapter.

Conflicting Evidence from Tax Disputes

While Republicans' intraparty disputes over the debt ceiling make spatial sense when we consider the difference between a centrist status quo point and an ideologically extreme reversion point, the history of roll call votes on tax policy from 2011 through 2016 breaks from this pattern in important ways. In 2001, President George W. Bush signed into law a set of tax cuts that would expire after 10 years. Had the votes been available, Republicans would have enacted tax cuts without an expiration date, but such a proposal would have been filibustered in the Senate, and Republicans didn't have the votes to invoke cloture. However, Senate rules automatically limited the amount of time allocated for debate on bills considered under "budget reconciliation" rules. The restriction was that bills considered under those rules could not increase the deficit for more than 10 years. So by setting the tax cuts to expire after 10 years, Republicans could bring their bill to the Senate floor under budget reconciliation rules and pass it with a simple majority. Their hope was that the tax cuts would later be either extended or made permanent because failure to do so would be functionally a tax increase.

Before their expiration in 2011, the cuts were extended for two more years, setting them to expire on January 1, 2013, unless extended again by the lame-duck session concluding the 112th Congress. Few members of either party wanted all tax cuts to expire. Cutting taxes has been the central, unifying purpose of the Republican Party since the Reagan administration, and while most Democrats wanted the tax cuts on the top income bracket to expire, most didn't want the rest to expire because of the continued weakness of the economy. Keynesian principles suggest that increasing taxes in a weak economy, particularly on those with the highest marginal propensity to consume, would be economically damaging. President Obama's position, shared by most congressional Democrats, was that the cuts should be made permanent on income up to $250,000 per year. The area of disagreement, then, was on income above $250,000 per year. As the deadline approached without a deal, Speaker Boehner developed what came to be called Plan B, which would be for the House to pass a bill permanently extending the tax cuts on income up to $1,000,000 per year. Then, if a deal were not reached before the deadline, Republicans would insist that the Democratic-majority Senate pass the House's bill and that the President sign it. That way, Republicans might avoid blame if negotiations failed, resulting in across-the-board tax increases. That would give Republicans a credible threat to walk away from negotiations, allowing them to extract more concessions.

However, getting the Republican House to vote for Plan B would not be easy. It would require convincing legislators to vote for a bill for which they might be *accused* of raising taxes, since some viewed a vote for any bill that did not extend *all* tax cuts as a vote to raise taxes, even though failure to pass a bill would actually result in *everyone's* taxes going up. The $1,000,000 threshold was actually far higher than what most voters wanted, but Republicans weren't worried about the consequences for their prospects in the general election. They were worried that if they voted for Plan B, they would face primary challenges for supposedly voting to increase taxes. To purists (i.e., Tea Party activists), it didn't matter on whom.

Moreover, since the purpose of passing Plan B was not to set policy, but to strengthen the party's bargaining position, the concern was a position-taking one. Why cast a vote that might risk a primary challenge when even the passage of the bill would have no direct policy effect? The model in chapter 4 is one in which legislators use party discipline to force themselves to cast electorally difficult votes in order to achieve policy outcomes. Thus, even the recipients of threats *want* that system in place. However, since Plan B was not intended to become law, nobody had anything to gain directly by passing it. Members of Congress just risked electoral backlash in the primaries. Perhaps, then, we should not be surprised that Speaker Boehner was unable to muster the votes to pass Plan B. Instead, he pulled the bill from the floor before a vote. Ultimately, a compromise was reached, making permanent all tax cuts on income up to $400,000, but while it passed by an overwhelming bipartisan majority in the Senate, it garnered most of its support in the House from Democrats, another Hastert rule violation similar to the October 2013 debt ceiling increase.

There are several obvious similarities between this story and the saga of the debt ceiling. Most importantly, many Republicans would have preferred to have a bill pass rather than to have no bill pass, but they felt electoral pressure, due to the perceived threat of a primary challenge, to vote against any deal. The result was that Speaker Boehner was forced to pass a bill that was closer to the preferences of most Republicans than the reversion point (of universal tax increases), but to do so primarily with Democratic votes, thereby violating the Hastert rule. However, there is a critical difference between the tax cut debate and the debt ceiling—the location of the reversion point.

With the debt ceiling, the reversion point (a debt ceiling breach and default) was an extremely conservative location, so only Republican extremists like

Representative Yoho sincerely preferred it to any given alternative. With respect to extending the 2001 tax cuts, though, the actual reversion point would have been a return to the higher tax rates set under the terms of the 1993 Omnibus Budget Reconciliation Act. That was not a conservative location. If anything, it was too far to the left for most *Democrats* at the time, suggesting that the midpoint between that reversion point and the eventual compromise should have put all or nearly all Republicans on the side of the alternative rather than the reversion point. And yet on the final deal, most House Republicans, particularly the most extreme conservatives, voted for the extremely liberal reversion point. How can we explain this?

One possibility is a more direct application of Groseclose and Milyo (2010). As with the debt ceiling, the strategic dilemma for Republicans was that the status quo point was not the same as the reversion point. In the absence of any bill passing, taxes would not remain fixed at their post-2001 levels. They would revert to their 1993 levels. So the order of policies along the liberal-conservative dimension was as follows: reversion < alternative < status quo. However, if Republican primary voters did not understand that the choice is really between the reversion point and the alternative point rather than between the alternative and the status quo, then they might treat a vote for the alternative as a vote for the more liberal rather than the more conservative option. Thus, Republicans would have to decide whether to cast what Groseclose and Milyo call a "position-taking" no vote, or an "outcome"-based yes vote in a circumstance in which constituents share the "outcome" preferences of their representatives, but fail to understand the strategic process. That would create precisely the type of collective action problem Boehner seemed to face.

However, there is another explanation. Gilmour (1995) observes that parties can have a strategic incentive not to accept a compromise if they believe that doing so will take away an issue that will be electorally beneficial, prevent getting a better policy later, or both. Republicans debating whether or not to accept a compromise by extending only part of the 2001 tax cuts had to ask themselves two questions. First, what would be the electoral consequence of a compromise, and, second, what would a subsequent Congress be able to accomplish?

Consider the following pair of scenarios. President Obama and House Republicans are unable to agree to a deal over partial extension of the 2001 tax cuts. Consequently, taxes revert to their 1993 rates. Republicans then blame President Obama and the Democratic Party for a tax increase and ride that issue to subsequent electoral victory, giving Republicans sufficient power to restore all of the 2001 tax cuts. On the other hand, in the second scenario, if Republicans agree to the deal, and more dangerously, *vote* for the deal, then they can no longer use a middle-class tax increase as an electoral weapon, nor can they criticize the deal without being subject to claims of hypocrisy. Moreover, even if they win the next election, they cannot simply restore the rest of the tax cuts because doing so would mean passing a tax cut *exclusively* on income above $400,000. If that is an accurate characterization of the dilemma, then Republicans voting against the deal were not really voting for the 1993 tax rates, but for the 2001 tax rates.

If we accept this characterization of the dilemma, then voting against the compromise actually leads to a full reinstatement of the Bush tax cuts after the next election. Thus, perhaps the real reversion point is not the 1993 tax system, but the 2001 tax system. However, if Republicans reject that argument, then the reversion point is the 1993 tax system. Rationalizing Republican Party divisions on tax legislation, then, may simply require understanding that legislators may differ not only in their policy preferences, but in their strategic assessments of where policy eventually reverts if nothing passes immediately.

However, this explanation also leaves something to be desired. First, chapter 2 rejected this type of rationalization for strategic voting among the electorate based on the observation that we cannot simply add stages to the model until behavior that initially appears irrational suddenly yields a positive outcome, and then stop the game. Doing so comes dangerously close to losing a coin toss and demanding "best two out of three." The even more complicated issue here, though, is that spatial theory itself becomes a questionable technique when different actors have different beliefs about the locations of the alternatives between which they are choosing.

However, if the logic of Gilmour's "strategic disagreement" reflects the substantive thought processes of congressional Republicans, then perhaps we should accept the need to vary the locations of reversion points—or at least perceptions of them—in spatial models. Moreover, this suggests that the divisions within the House Republican caucus since 2010, revealed by the politics of reversion points, run even deeper than one might suspect, encompassing both policy preferences and strategic perceptions.

The Future of Intraparty Divisions and Party Leadership in Congress

The conventional explanation for intraparty divisions within the Republican Party in the post-2010 era is that the rise of the Tea Party and the threat of primary challenges have driven a wedge between the establishment wing of the party and the more fractious, extremist wing currently embodied by the House Freedom Caucus. However, the argument presented here is that the divisions that arose after 2010 are more complex, and resulted as much from changes to the legislative agenda that occurred due to a combination of extreme polarization and divided government. If the divided government legislative agenda was as responsible as primary threats, then what happens with a return to unified government?

The 2016 election saw the surprise victory of Donald Trump at the presidential level, along with the continued Republican majorities in the House and Senate. What kind of relationship should we expect to see between a Speaker and a similarly composed Republican House caucus in this context? The popular explanation for divisions among congressional Republicans is that they stem from a caucus that shifts continually rightward in response to the threat of primary challenges. If that were the whole story, then a return to unified government would change nothing about the tension between Republican leaders and their rank-and-file. However, the analysis here suggests otherwise. Understanding what should happen under unified Republican control requires addressing several points.

First, what will be the contours of the legislative agenda? The items that have split the Republican caucus in recent years still need to be addressed. Congress needs to fund the government, which requires either appropriations bills or continuing resolutions. Furthermore, without a balanced budget, Congress still needs to raise the debt ceiling, and even the budgets proposed earlier by now-Speaker Paul Ryan continue to run deficits for an extended period of time. Hence, the bills with ideologically extreme reversion points remain on the legislative agenda. However, the dynamic that has split the Republican Party in response to these agenda items does not. After 2010, Republican leaders tried to manage these bills by using them as "hostages," in Senator Mitch McConnell's words, with which to extract concessions from a Democratic-majority Senate and a Democratic President. That allowed Republicans to use these concessions as carrots to attract enough Republican votes to avoid the reversion points, while simultaneously telling conservative activists that they don't actually *want* to raise the debt ceiling—they just do it because it is the only way to extract spending cuts from Democrats. Without a Democratic president or majority in the Senate, the hostage-taking approach cannot work.

With respect to the debt ceiling, there are few options since the debt ceiling simply must be raised, and Republican leaders know this. The easiest approach would be something like a return to the "Gephardt rule," named for Rep. Richard Gephardt's (D-MO) tactic of attaching debt ceiling increases to spending bills under a self-executing rule, but if the Republican Party is unwilling to vote for "clean" debt ceiling increases, then it must attach debt ceiling increases to bills that they will otherwise pass as a party. The most strategically sound approach on general fiscal matters would be to simply pass ideologically extreme fiscal measures using budget reconciliation on strict party-line votes, which cleanly divide the party caucuses, invoking the principles of chapter 2. Since the party cannot afford a shutdown during unified government, it cannot afford to give any bargaining leverage to Democrats, suggesting the use of budget reconciliation to block Democratic filibusters in the Senate. That would allow fiscal legislation to move as far right as is necessary to satisfy the House Freedom Caucus and other intransigent conservatives, thereby preventing the conflicts that plagued Boehner's term as Speaker. As long as fiscal legislation can move rightward, there is less motivation for the intraparty conflict we observed from 2011 through 2016, and unified government provides procedural options for that. Thus, while Congress must continue to address the types of bills that involve ideologically extreme reversion points, there is less reason for a Republican Speaker to continue pushing cross-pressured caucus members into coalitions with the minority party during unified government. The easier approach is for policy to move rightward in order to satisfy the wing of the party that had formerly been at odds with leadership. The House Freedom Caucus spent John Boehner's term in office at odds with him, but Speaker Ryan has every incentive to cater to their preferences.

Moreover, if it is the combination of extreme polarization and divided government that reduces incentives for the House to take up "normal" legislation, then the return of unified control gives the House incentives, once again, to take up bills that move policy from a fixed status quo to some specific alternative. Such bills do not pit Republicans against each other in the same way that debt ceiling votes do because they don't involve ideologically extreme reversion points. Rather, they unify the caucus against moderate-to-liberal status quo points, thus giving us more examples of bills on which Republicans' ideological preferences are unified. What happens on these bills, then, depends on whether or not Republicans continue to perceive ideologically conservative votes as electorally optimal because of a persistent fear of primaries.

If Republicans' fears of Tea Party primary challenges are real and remain a stable force in American politics, then when a Republican majority takes up "normal" legislation, floor votes would not be subject to a collective action problem of sincere voting. Republicans' policy preferences and *perceived* electoral pressure would both push them to vote for conservative alternative points rather than centrist status quo points. For all practical purposes, the caucus would be both ideologically and electorally homogeneous, allowing the caucus to achieve policy goals through sincere roll call voting without any party pressure. Party leadership would be superfluous, and the model would revert to essentially a Krehbiel (1998) model of the legislative process in which policy outcomes are determined by the distribution of ideal points and pivot points. What we would *not* observe is any attempt by majority party leadership to push caucus members into coalitions with the minority party amid violations of the Hastert rule, as Speaker Boehner repeatedly did.

Alternatively, suppose that the phenomenon of the Tea Party primary challenge was a temporary historical curiosity. If Republicans' fears of primary challenges dissipate under unified Republican control, then the model simply reverts to the process described in chapter 4, in which case parties act as preference-preserving institutions to solve the collective action problem of sincere voting by counteracting the electoral pressure to cast insincere "centrist" votes. Again, what we would *not* see is Republican leadership attempting to fracture rather than unify the caucus.

The key point is that without divided government, there is less need for Republicans to make concessions to Democrats, and hence fewer fractures within the party over whether or not the party has conceded too much in order to avoid ideologically extreme reversion points that some caucus members don't really mind anyway. Consequently, the Speaker has less need to cobble together bipartisan coalitions that violate the Hastert rule. The unusual intraparty divisions recently observed within the Republican Party, then, will likely be mitigated by the return to unified control.

Mitigated, though, does not mean eliminated. When the 2010 election produced a combination of divided government and extreme polarization, the resulting legislative agenda revealed previously hidden ideological divisions, and those divisions still exist, along with their attendant complications and animosities. Those animosities didn't go away just because the Republicans gained unified control of government, and their effects remain uncertain.

From the vantage point of 2017, the scandal-plagued Presidency of Donald Trump and associated political dysfunction may complicate tests of any model, but the games of budgetary brinksmanship that pitted Speaker Boehner against the extremist faction of his own party have not appeared as contentious. The Republican Party seems to understand that it cannot avoid blame for a government shut down or debt ceiling breach under unified government, so threats to do either have diminished, and efforts to avoid either have intensified. The legislative agenda has also expanded, including notably to the eventual passage in the House of the American Health Care Act of 2017, which was the House Republicans' plan to "repeal and replace" the PPACA. While the process of passing the AHCA through the House was more difficult than the symbolic "Obamacare repeal" votes that the House repeatedly held when Republicans knew those bills had no chance of passage, that is actually the point. The AHCA was intended to become law, even though it failed in the Senate. The point, though, is that the legislative agenda has expanded in such a way that it is no longer merely about avoiding reversion points. Consequently, it now unifies the party against the Democrats, indicating a return to more normal legislative coalition politics. Many of the House Republicans representing more moderate districts were nervous about voting for the AHCA and were subject to party whipping, reflecting a return to the process described in chapter 4. Again, that suggests the politics of 2011 through 2016 resulted, not merely from primary challenges and the threat they posed, but from the changes to the legislative agenda brought on by the combination of extreme polarization and divided government. Speaker Ryan remains at odds, in many ways, with the most extreme wing of his party, who complicated the negotiations, and it is important to remember that the ideological divisions revealed by the agenda from 2011 through 2016 are real and didn't disappear. They *can* be hidden or mitigated by the structure of the agenda, but those disagreements exist and will continue to exist. A substantive political question, then, is the extent to which their past revelation affects future political negotiations simply on the basis of distrust. However, from the vantage point of 2017, current tensions are less than those between Speaker Boehner and the Tea Party in, for example, 2013, as we would expect if those tensions were a product of the legislative agenda.

Critically, then, the principles of the unified model allow us to make predictions about the future of the Republican Party that differ significantly from conventional explanations. If the intraparty divisions among Republicans merely reflect their newfound fear of primary challenges, as conventional wisdom holds, then regardless of the return to unified control, the divisions should remain as long as the fear of primaries remains. However, if the intraparty divisions resulted from the diminished legislative agenda that we observed as a consequence of combining extreme polarization with divided government, then regardless of whether or not Republicans continue to fear primaries more than general elections, the return to unified government should mitigate intraparty divisions by suppressing conflicts over bills that involve ideologically extreme reversion points and refocusing the agenda on items that involve centrist status quo points, which unite the party more easily. In any case, understanding these divisions requires understanding how parties manage the roll call voting process in response to the collective action problem of sincere voting, which remains the central issue for the incremental positioning process.

8

Conclusions

This book began with a set of puzzles. The modern Congress is virtually defined by some combination of ideological and partisan polarization. Why do legislators adopt such extreme positions? How do they win given their extremism? What role do parties play in creating or enabling that extremism? Most of the advances in spatial theory for the last five decades have been piecemeal efforts to address some aspect of these puzzles, but many of these advances have explained polarization by diluting the effect of spatial reasoning within the model. This book has taken an alternative, and more holistic, approach by presenting a unified spatial theory of legislative elections, parties, and roll call votes. The model has linked the two strands of spatial theory—spatial models of elections and spatial models of roll call voting—and has done so by incorporating parties into the model in a way that reconciles some of the apparent conflicts between partisan theories of Congress.

The key to the model was to reverse the order of events from conventional spatial models of elections. The incremental positioning model begins, not with a round of legislative elections, but with a legislative session. In the first stage, a majority party caucus selects a party/chamber leader, whose platform consists of a disciplinary regime. The winner of that contest carries out her platform by managing the legislative session. Throughout that session, each legislator incrementally adopts a position in the policy space by casting roll call votes, constrained by the disciplinary regime. Then, each incumbent faces a re-election campaign in which voters evaluate incumbents, based not on their incrementally adopted locations in the policy space, but on the bundles of roll call votes used to construct those locations.

The predictions of the model closely resemble modern legislative politics, explaining several empirical puzzles while reconciling existing theories. When a legislative caucus is ideologically homogeneous, electorally diverse, and policy motivated, it will select a leader who promises to apply "preference-preserving influence," meaning that she will threaten to impose modest punishments on legislators who bow to electoral pressure to vote insincerely, thereby solving the collective action problem of sincere voting. Consequently, legislators with noncentrist policy preferences who represent centrist districts will vote sincerely, thereby incrementally adopting ideologically extreme and electorally suboptimal positions in the policy space. However, the structure of the legislative agenda will reduce the electoral costs of that extremism by creating a "proximity paradox," in which some voters prefer the bundles of roll call votes implied by distant locations in the policy space to the bundles implied by more proximate locations. In the end, these extreme incumbents will receive lower vote shares than they would have received had they incrementally adopted more centrist locations, but they offset that loss of utility with policy victories that they could not have achieved without parties to apply preference-preserving influence, and they generally manage to win re-election anyway with only somewhat lower vote shares because of the difference between how rational voters respond to candidate locations and how they respond to the bundles of roll call votes implied by those locations.

Thus the ideologically extreme and electorally suboptimal positions that legislators take in the modern Congress are rational and compatible with an undiluted form of spatial theory. Polarization and its electoral consequences are easily compatible with spatial theory as long as we link spatial models of elections to spatial models of roll call voting through party leadership in Congress in an incremental positioning model.

Moreover, even the unusual patterns that developed in the House of Representatives after the 2010 election are illuminated by the incremental positioning model. While conventional wisdom holds that the intraparty tensions among House Republicans that emerged after 2010 result from fear of primary challenges, the incremental positioning model demonstrates that the unusual divisions between former Speaker John Boehner and the radical wing of his caucus that deposed him follow naturally from a subtle agenda shift. The combination of extreme polarization and divided government produced by the 2010 election stripped the legislative agenda of nearly everything except for the bills with ideologically extreme "reversion" points, such as breaching the debt ceiling. Such bills divided the Republican caucus in ways that make sense within the context of the unified model. The disputes that began to plague the Republican Party after the 2010 election may have existed beneath the surface prior to 2010, but it took an agenda shift to reveal those divisions, as demonstrated by the model.

Of course, the model is not a complete explanation for legislative polarization. Rather it is a way to reconcile polarization with spatial theory, and accordingly a variety of puzzles remain. Three, in particular, stand out. First, the model explains why legislators with extreme preferences express those preferences through their roll call votes even when doing so seems irrational, but it does not explain why legislators have noncentrist policy preferences in the first place. Second, the model has focused on the House of Representatives, but the Senate, too, has become more polarized despite weaker disciplinary tools for Senate leaders. Why? Finally, going forward, how stable will the unusual developments in the House be?

Remaining Puzzles

In a caucus characterized by ideological homogeneity but electoral heterogeneity, polarization results from party influence. The persistence of electoral heterogeneity is easy to explain given existing research on redistricting. While many commentators hold the mistaken belief that sequential gerrymanders have eliminated competitive districts, the competing incentives faced by those drawing district lines, as described by Butler and Cain (1992), explain why electoral heterogeneity persists. The greater mystery is why a caucus becomes more ideologically homogeneous over time.

We know that as the Democratic Party adopted civil rights legislation into its platform, the southern Democrats who had been part of the party coalition since the Civil War began to switch parties, or retire to be replaced by Republicans. As they did so, the Democratic caucus became more homogeneous simply because one of the two factions left the party, leaving only the northern liberals. At the same time, the moderate contingent of the Republican Party, which had been based in the Northeast, began to disappear, to be replaced by northern Democrats who were at least as liberal as the existing members of their caucus. The new southern Republicans, in contrast, were quite conservative as a result of the coalitional forces that created the southern Republican Party (Polsby 2004). Historical analysis, then, explains *part* of how the two caucuses became more homogeneous, but there are two problems with settling for an explanation based on the history of political geography. First, such analysis doesn't yield any generalizable theory. Historical happenstance may be real, but it tells us little about how or why future changes may occur.

Even more puzzling is the fact that the party split has continued to grow in surprising ways even after the parties sorted themselves out across the Mason-Dixon Line. By the 1990s, few "Dixiecrats" remained in the South, and few northeastern Republicans remained. However, while congressional Democrats are not markedly more liberal now than they were in the 1990s, the Republican caucus has continued to move to the right, creating an asymmetric polarization evident in DW-NOMINATE scores. Consider Figure 8.1, showing the party medians in the House of Representatives from the 83th Congress through the 113th Congress, which has been the focus here.

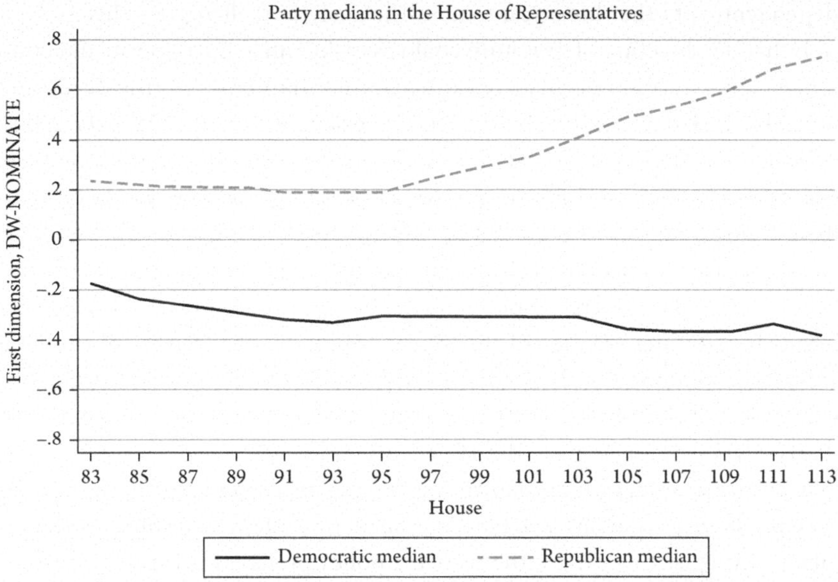

Figure 8.1 Party medians in the House of Representatives

So, while we can explain why legislators with noncentrist preferences express those preferences on roll call votes, it is more difficult to explain how we got this particular array of noncentrist preferences. Pundits typically place blame on the usual suspects, such as redistricting. As discussed in chapter 5, though, redistricting has very limited capacity to explain polarization, which is consistent with previous analysis (Abramowitz et al. 2006; Brunell 2006, 2008; Buchler 2005, 2009, 2010a, 2010b; McCarty et al. 2009). Unfortunately, other potential explanations are more difficult to link directly to the underlying, personal policy beliefs of members of Congress. We can, however, speculate.

The pool of viable candidates for the US Congress comes disproportionately from state legislatures. Masket (2009) argues that party activists polarize state legislatures, and by doing so they may indirectly polarize Congress by polarizing the potential candidates for Congress. In principle, a party apparatus may empower itself by recruiting the types of candidates who are likely to empower it in the legislature once elected. However, drawing direct connections between these observations becomes more difficult at each stage.

The underlying beliefs of party members may also be affected by aspects of the party network, such as the think tanks that began to proliferate in the 1960s. Perhaps organizations like the Heritage Foundation and the Cato Institute have succeeded in shaping the opinions of potential Republican candidates so effectively that the rightward march of the Republican legislative caucus continues independently of other historical trends. The analytic challenge for this explanation is that the direction of causation could as easily run in the opposite direction. Ideological polarization might increase demand for think tanks to provide research compatible with one's existing beliefs (see, for example, Bertelli & Wenger 2009).

Changes in the structure of the media may play a similar role. Between cable news channels and internet news sources, anyone who wants to avoid encountering opinions or even factual information that might conflict with her general predispositions can easily do so, and with decreasing exposure to alternative viewpoints, sincere opinions might become more polarized. However, without longitudinal studies of the sincere opinions of the pool of potential candidates, combined with their media preferences, this too is speculative. Nevertheless, it is probably worth noting that Congress was polarized in the era of party-owned newspapers in the 19th century.

Given these observations, explaining the polarization of legislators' sincere policy preferences remains a challenge. The incremental positioning model can explain why legislators *express* their sincere policy preferences through their roll call votes despite electoral pressure to do otherwise, but it provides no additional insight beyond historical analysis into why those underlying preferences change over time. In fact, a model with the legislative session occurring first is the wrong way to approach that particular question anyway because it takes legislators' policy preferences as initial conditions. More than anything else, then, this distinction points to the importance of using election-first models and legislative session-first models in conjunction with each other because each type of model explains something that the other cannot. Examining the process from an election-first perspective may help to explain how we wind up with legislators who have noncentrist policy preferences, but it cannot explain why those preferences are expressed through the roll call voting process. Examining the process from a legislative session-first perspective can explain why the roll call voting process plays out the way it does, but it cannot explain why legislators have noncentrist preferences at the beginning of the legislative session. Alternative model structures yield different and, in this case, complementary insights. That is why we need to avoid always using models structured the same way. Variation in model structure gives us variation in explanatory power.

Another set of questions arises regarding the Senate. This book has focused largely on the House of Representatives, where party leaders have at least somewhat stronger disciplinary tools and more control over the legislative agenda, both of which are central components of the model. However, despite the fact that Senate leaders have weaker tools, in terms of both discipline and the agenda, we have seen increasing polarization in the Senate as well, which is itself an important observation that contradicts the common argument that redistricting causes polarization.

Several possibilities are immediately obvious. First, recall from chapter 3 that if a legislator is pivotal, then an insincere roll call vote can be unboundedly costly. Thus, the cost of adopting an insincere position in the policy space can be unboundedly costly. If noncentrists expect to be pivotal, then, they should incrementally adopt noncentrist locations in the policy space with or without parties because the marginal electoral cost of a single sincere roll call vote will be quite small, compared to the potentially unboundedly large policy cost of an insincere roll call vote. In the House, this is not a compelling explanation for polarization because so few roll call votes are decided by a single-vote margin. However, the Senate is a much smaller body, so any individual senator has a higher probability of casting a pivotal roll call vote. Moreover, while the main pivot point in the House is the 218th vote, there are two main pivot points in the Senate—the 51st vote for passage and the 60th vote for cloture (in addition to veto override points in each chamber, which are rarely invoked).

Arguably the most significant piece of legislation considered by Congress in decades was the Patient Protection and Affordable Care Act, passed in 2010. In the Senate, Democrats held 60 seats when the bill first reached the floor, and the Democratic caucus invoked cloture on a strict party-line vote. Every Democratic senator, then, cast a pivotal vote on the most significant piece of legislation in decades. By virtue of a smaller body and another pivot point, senators are more likely to cast pivotal votes than House members, so by the reasoning in chapter 3, parties may be less necessary for sincere voting, and hence less necessary for polarization in the Senate. If so, then the comparatively weaker tools that Senate leaders have for structuring the agenda and imposing discipline may simply not play as significant a role.

Recall, also, that the disciplinary tools required to apply preference-preserving influence may be quite weak—potentially as weak as simple "demerits" or even social pressure. If so, then shifting our attention from preference-deviating influence to preference-preserving influence may explain how the Senate could display patterns similar to the House when leaders have even weaker disciplinary tools available to them.

Regardless, explaining polarization in the Senate is a more challenging endeavor. However, the process by which senators adopt extreme positions in the policy space is the same as in the House—the incremental process of casting roll call votes, and if voters are rational, they will respond to Senate voting records the same way they respond to House voting records. Thus, even if the specifics of the model proposed in this book are more closely related to the House of Representatives than to the Senate, the underlying methodological point about the construction of spatial models is as applicable to the Senate as to the House.

The final question is about how stable the past dynamic between Republican leaders and their rank-and-file will be. As chapter 7 discussed, Republican legislators are incrementally adopting ideologically extreme positions in the policy space, but currently they do not appear to be doing so by empowering party leaders to enforce sincere voting. If anything, Republican leaders since the 112th Congress have been trying to use party discipline to *temper* the extremism of the caucus. Chapter 7 argued that the dynamic from 2011 through 2016 resulted not merely from the emergence of primary threats as the predominant electoral concern among caucus members, but from a House unable to pass normal legislation due to the combination of extreme polarization and divided government. In that context, all that remained on the legislative agenda was a set of bills needed to avoid ideologically extreme reversion points, such as a debt ceiling breach or a government shutdown, and these agenda items highlighted previously hidden divisions within the caucus. As we enter a period of unified government at the time of writing, the question is about what happens next. The divisions between Republican leaders and their extremist faction are real, and revealed by the prior legislative agenda, but if that agenda was the result of combining extreme polarization with divided government, then a return to unified government should mean a return to the kinds of agendas that unify rather than divide the majority caucus. The Freedom Caucus may be a distinct faction from Republican leadership, but if distinctions are hidden or revealed by legislative agendas and political circumstances dictate the agenda, then tensions should at least diminish somewhat under unified government, providing a partial test of whether the tensions between Boehner and the extremist faction were merely about primaries or based significantly on agenda changes. Time will tell. To the degree that party leaders want to influence roll call voting patterns, though, the central issue remains the collective action problem built into roll call voting. Moreover, any attempt to enforce discipline must be built around rewards and punishments with magnitudes that are calibrated to the incremental effects of each individual roll call vote.

Implications for Future Research

The unified model leaves us with three sets of critical issues relevant to future research: the relationship between spatial theory and specific institutional rules, the observational consequences of party influence, and its normative implications.

Spatial Methods and Institutional Structures

The models posed by Downs (1957) to explain electoral politics and by Black (1958) to explain the process by which groups select policy choices have been invaluable to the study of legislative politics. Whatever flaws there may have been in these models, they have generated a treasure trove of research. However, there is a limit to what spatial models of elections or spatial models of committee decisions can achieve in abstraction or isolation.

The model posed by Downs to explain electoral politics was a marvel of parsimony, but there is a direct trade-off between generalizability and explanatory power. Suppose that two office-seeking parties simultaneously select locations in the policy space to compete for votes. The winner then sets policy. The aspect of this model that makes it appear generalizable is that it does not attempt to characterize the specific actions taken by the winner of the election at the end of the game, so in theory it should apply equally well to any type of election. Yet that is also its weakness. Choosing between two policy locations and choosing between the bundles of roll call votes associated with those locations are both conceptually and computationally different, and that difference can affect preference-orderings. So, in fact, the lack of specificity regarding the policymaking process *limits* the applicability of the Downsian model. The mathematics of our models must follow from the substance of the political process, not computational convenience.

If the winner of the election can take unilateral action to set policy, then a single-stage platform implementation process makes substantive sense. However, the more the office in question is subject to institutional constraints that prevent the winner of the election from unilaterally setting policy, the less useful that type of model will be. Legislators face a variety of institutional constraints, from the legislative agenda to party discipline, yet those constraints are the result of strategic choices by legislators. That is why rational legislators will sometimes adopt electorally suboptimal positions, and why voters re-elect them anyway. We cannot construct spatial models that abstract too far away from actual political processes.

In fact, the principle of incrementalism, central to the unified model here, may provide different insights in different legislatures. Any legislative election involves a choice between bundles of roll call votes, but other factors must be addressed. Consider the distinction between the US Congress and PR-based legislatures. While legislators in the US Congress and members of parliament in systems with proportional representation all cast roll call votes, thereby incrementally adopting locations in the policy space, two critical differences emerge that give the unified model from chapter 4 limited applicability to PR systems. First, a pure PR system is one in which all legislators run nationwide, meaning that they all represent the same geographic constituency. Thus, there is no role for electoral heterogeneity, which is a critical condition for party government in the unified model. Second, since individual legislators are granted a position on the party slate at the discretion of party leadership, they have limited capacity to vote against party leadership, meaning strong rather than weak party discipline.

The unified model was developed to explain the process in legislatures with weak party discipline, rather than systems such as PR. Thus, it also has less applicability to a legislature using the Westminster system, in which legislators are similarly dependent on party leadership for their places on the ballot. The key lesson for such legislatures, then, is in how *parties* construct bundles of policy positions in order to invoke the proximity paradox, derived in chapter 2. That remains applicable. Beyond that, spatial models constructed to explain any *specific* legislature must take into consideration the specific institutional rules of that legislature.

More than anything else, then, the methodological observation of this book is that spatial models cannot be constructed in a manner that is blind to the institutions for which candidates are running. In the context of legislative elections, this is an easy piece of advice to follow since spatial models of elections and spatial models of legislative roll call voting are so structurally similar that they can be linked in a unified model with relatively few mathematical complications. Can spatial models of elections be used to explain executive elections, judicial elections, or even elections for other types of administrative positions? Only if the actions taken by these officers can be characterized plausibly by spatial theory. A one-size-fits-all model doesn't really fit any particular size.

Even beyond the manner in which spatial models operationalize the policymaking process, the basic structure of these models limits the type of questions that can be asked. Broadening the types of structures we use, then, expands the realm of inquiry. From the perspective of traditional spatial theory, the most unusual feature of the incremental positioning model derived here is not how it characterizes the platform implementation process. That is taken wholesale from spatial models of roll call voting. The most unusual feature is that the election occurs after the legislative session rather than before it. There is no obvious reason that the election must always come first, unless we are attempting to model the entire history of Congress dating back to the 1788 election, because every round of legislative elections other than 1788 was also preceded by a full Congress. It is simply a *tradition* that we place the election first, and we can always take another perspective. Consider the nature of incumbency. Most legislative elections are contests between an incumbent and a challenger, and more than 90% of the time in House elections, the incumbent wins. From one perspective, we might argue that incumbent re-election rates suggest that we should place the election first in order to explain how the incumbent got into office in the first place, since every subsequent election is almost a foregone conclusion. From another perspective, perhaps we should place the legislative session first in order to explain what incumbents do in office to allow them to defeat their challengers despite the fact that they seem too ideologically extreme to win. Chicken and egg . . .

Fifty years of universal adherence to election-first tradition has blinded spatial theory to the opportunities created by a model built in reverse order. Placing the election first has drawn attention to the strategic interaction between opposing candidates operating within the same electoral context. Competitors offer alternative platforms in order to compete for votes, constrained by their own policy preferences, their opponents' strategic choices, and other contextual factors. Does this strategic interaction occur? Certainly, but it is not the only strategic interaction relevant to legislative elections. Equally important are the interactions between legislators operating across different electoral contexts, because what happens in the legislature has important implications for incumbents' re-election prospects. By placing the legislative session first, the incremental positioning model has drawn attention to the strategic interaction between members of a legislative caucus, representing different electoral constituencies, because party strength is determined by that strategic interaction. Moreover, party discipline has important electoral implications. Rather than looking for explanations for polarization in campaign dynamics, then, the incremental positioning model has looked for explanations in legislative dynamics. However, unlike spatial theories of legislative politics that exist divorced from elections, the model presented here has traced out the electoral implications of legislative dynamics by examining the implications of party politics for the spatial-electoral positioning process, and explained legislative politics as they occur in anticipation of the *re*-election campaigns that surprisingly many incumbents win.

It is *not* the argument of this book that models *must* place the legislative session first. Rather, the structure of our models determines the kinds of questions we can ask. Since campaign dynamics and legislative dynamics interact, we must use election-first models and legislative session-first models in conjunction with each other, and more generally, experiment with alternative model structures in order to broaden the range of questions we can ask.

Political Parties

Parties remain perhaps the most controversial institutions in spatial theory. Empirical scholars of Congress place parties at the center of their studies, while spatial theorists largely ignore them as merely the byproducts of underlying ideological divisions. The dispute comes from a misinterpretation of one of the central findings in empirical studies of roll call voting and legislative politics more broadly. Legislators' behavior and resulting policy outcomes can be explained largely by estimates of their locations in the policy space, with little additional predictive power coming from legislators' party affiliations. To party skeptics, this means that parties are merely the incidental byproducts of spatial politics. To party adherents, this means that the operation of parties is just more subtle than simple pressure and occurs primarily through procedural manipulation. Skeptics retort that an ideologically homogeneous majority caucus that does not impose procedural control will achieve similar policy outcomes anyway, and so the dispute continues.

By reformulating the conditions for party government, though, we have seen that competing theorists have been talking past each other. Among an ideologically homogeneous but electorally diverse caucus, sincere voting is a collective action problem, so the caucus will not vote cohesively in the absence of party influence. On the other hand, if such a caucus is policy motivated, then it will delegate power to party leadership to enforce sincere voting. By acting as "preference-preserving institutions," then, parties serve as the reason that legislators' sincere policy preferences determine their roll call votes. Observationally, though, this creates a challenge because party influence exists primarily to counteract other pressures.

To make another analogy, suppose we observe a stationary iron ball on a flat table. The fact that the ball is not moving *suggests* that the only forces acting on it are the precisely balanced forces of gravity and the force from the surface of the table. However, that is not necessarily the case. If there are two magnets exerting an equal amount of force in opposite directions laterally, the net effect will be that the ball remains stationary. So suppose that we observe a strong magnet on the left side of the ball. Since the ball is stationary, there must be some other force acting on the ball pulling it to the right with precisely the same amount of force. Occam's razor suggests another magnet, although stage magicians could devise more creative methods. Regardless, the ball's stationarity is not a result of the absence of lateral forces acting on it. The ball is stationary because the horizontal forces acting on it are canceling each other out. Moreover, if we *observe* one such force, we must *infer* the existence of at least one more such that the net effect of all forces acting on the ball is zero.

We observe members of Congress voting in ways that appear mostly sincere given the lack of dramatic movement when they decide to retire. We observe electoral pressure acting on members of Congress to vote insincerely in order to incrementally adopt more moderate positions in the policy space. We observe them casting electorally suboptimal roll call votes despite that pressure, even when those votes are not pivotal. The fact that they do so suggests that something is counteracting electoral pressure toward centrism. Strong policy preferences aren't enough because we observe electorally suboptimal roll call votes even when legislators are not pivotal. Pressure from primaries helps to explain Republican dynamics after 2010, even if it is an overreaction to a few anecdotes, but the patterns are more broad than that. Interest groups do not appear to have sufficient strength to achieve the effect. Rather, even the legislators from the most marginal districts seem to track their party medians over time, strongly implying that it is parties who counteract the electoral pressure toward centrism, which is exactly what we would expect when the conditions for party government are met.

This creates a strange tension between the subtlety of the empirical patterns by which we can infer party influence and the crudeness of the influence itself. The empirical patterns that suggest party influence are somewhat indirect because the nature of the effect is to cancel out other pressures. However, the mechanism is relatively crude—direct pressure on substantive roll call votes. While procedural control, manipulation of the committee system, and other subtle forms of party influence are important, the argument in this book is that it is difficult to explain polarization on substantive policy votes without blunt power as well. Moreover, this blunt application of power is a manifestation of the preferences of the rank-and-file rather than a corruption of those preferences. Members of Congress *want*, collectively, to be pressured by party leaders. It is just that they *want* to be pressured to do what they would do in the absence of *other* constraints because they are *not* acting in the absence of other constraints. The competing pressures acting on members of Congress are not independent forces—they arise in response to each other. Smith (2007) argues that the challenge of finding party influence in Congress is that it cannot be measured directly. The model here explains why.

Parties and Representation: Betcha Can't Shirk Just Once

If parties affect the legislative process by canceling out electoral pressure toward centrism, as the incremental positioning model suggests, that has important implications for representation. As H. L. Mencken wrote, "Democracy is the theory that the common people know what they want, and deserve to get it, good and hard." If legislators empower parties to *prevent* themselves from giving the common people what they want, then parties undercut responsiveness. Chapter 4 drew an analogy between encouraging legislators to disregard electoral pressure, one vote at a time, and the metaphorical devil on one's shoulder who argues that one more potato chip won't really matter. Except for the occasional high-profile vote, such as the 1993 Omnibus Budget Reconciliation Act or the 2010 Patient Protection and Affordable Care Act, each individual roll call vote used to construct a sincere but ideologically extreme voting record will have a negligible impact on an incumbent's vote share, but the cumulative effect of repeatedly voting in an ideologically extreme manner may be significant. Similarly, each individual potato chip will have a negligible effect on one's weight, but eating chip after chip based on estimates of the marginal utility of the next chip and the marginal effect of that chip on one's waistline may lead to an ever-expanding waistline. Implicit in that analogy is a value judgment. We associate overindulgence with moral failure, and we associate policy nonresponsiveness with democratic failure, at least if we adopt a delegate model of representation rather than a trustee model. Parties enable policy-seeking legislators to shirk in their delegate duties by altering the calculus of positioning at an incremental level so that legislators are willing to commit a series of minor infractions that they might not commit if these infractions had to be committed in one grand act of representational defiance. One more potato chip won't really matter anyway, right? So says the metaphorical devil on our shoulder.

However, to argue that party-facilitated polarization is *unrepresentative* is not to argue that party-facilitated polarization is intrinsically *bad*. Many pundits worship at the altar of centrism as though politico-moral virtue were defined by the arithmetic midpoint between the two parties' platforms. But that midpoint may not translate to a strong economy, national security, or any other desirable outcome, nor even a healthy political system.

In fact, it was not too long ago in historical terms that political scientists advocated something like polarization. When the American Political Science Association published *Towards a More Responsible Two-Party System* in 1950, it was during a period in which many Democratic and Republican politicians had overlapping policy platforms. Yet many scholars longed for a system in which Democratic politicians consistently advocated one set of policies, and Republicans another. That way, party labels would serve as clear signals about candidates' policy platforms, and voters could ascribe credit or blame for national conditions to one party when its platform was enacted. Polarization means we now have something closer to what the 1950 APSA committee recommended.

While it is easy to bemoan polarization, then, and to criticize parties for undercutting representation, we should remember that the alternative simply creates different problems. Moreover, since there are no coherent arguments that the arithmetic midpoint between the Democratic and Republican platforms strengthens either the economy or national security, we should avoid pining for the bygone era of centrism. If one of the parties' platforms actually would strengthen the economy and ensure national security, then isn't it better that at least one party consistently advocates such a platform, regardless of where the midpoint is?

Appendix A

INCREMENTAL POSITIONING AND MULTIDIMENSIONAL POLICY SPACES

The unified spatial model presented here has examined why candidates adopt extreme positions and how they win in a unidimensional policy space. Chapter 2 demonstrated that voters with single-peaked, symmetric preferences over policy can paradoxically prefer a distant candidate to a more proximate candidate for legislative office because the bundle of roll call votes associated with the former can provide higher utility than the bundle associated with the latter. Chapter 2 posed this observation as an explanation for the failure of the median voter theorem. However, chapter 3 pointed out that as long as legislators aren't pivotal, they can position themselves costlessly anywhere in the policy space, and the electorally optimal position remained the location of the median voter. Thus, sincere voting is a collective action problem, and chapter 4 argued that parties solve that collective action problem by incentivizing sincere voting when the conditions for party government are met.

All of this, though, is predicated on a unidimensional spatial model. The puzzles the book addresses are essentially failures of the median voter theorem, and the median voter theorem only operates in a unidimensional policy space, because if the policy space is multidimensional, then there is no median voter. Rather, as McKelvey (1976) and Schofield (1978) have observed, the intransitivities created by a multidimensional policy space allow a variety of perverse outcomes, and even in the modern Congress, with the first dimension of NOMINATE explaining more than 90% of roll call votes, the political world is never purely unidimensional. All it takes is a slight deviation from unidimensionality and the median voter theorem collapses.

This raises two questions about the applicability of the unified model presented in this book. First, is there even a puzzle to address, and, second, can the principles of the unified model work in a multidimensional policy space? The answers are yes and yes.

Multidimensionality and the Puzzle of Polarization

First, consider the electoral success of ideologically extreme incumbents. While the policy space in legislative politics is not unidimensional, Poole and Rosenthal (1997) demonstrate that in the modern era, something like the liberal-conservative spectrum, approximated by the first dimension of NOMINATE, explains most of roll call voting. Within that dimension, most incumbents incrementally adopt positions far from 0, and yet win re-election. But, if the policy space is not unidimensional, then perhaps they win re-election under simple proximity voting by exploiting the intransitivities created by additional policy dimensions. If so, then there is no puzzle to explain. The median voter theorem simply fails because there is no median voter. While mathematically possible, is this substantively plausible?

Consider an n-dimensional policy space. For demonstration purposes, consider a single voter with a quadratic loss utility function and an ideal point at the origin $(0, 0, \ldots, 0)$, where the first dimension is the liberal-conservative dimension. Suppose that an incumbent has an "extreme" score within the first dimension of $r > 0$. Suppose, further, that the challenger has a less extreme score within that dimension of $d < 0$ where $|d| < r$. What would have to be true for the voter to prefer the incumbent under standard spatial assumptions? Geometrically, since the challenger is closer to the voter along dimension 1, she must be farther away in at least one other dimension. Let us simplify the process by setting the incumbent's location to $(r, 0, 0, \ldots, 0)$.

If the challenger's location is given by the vector $D = (d_1, d_2, \ldots d_n)$, then the condition is simple. The voter will prefer the "extreme" incumbent to the challenger if $D \bullet D < r^2$. In two dimensions, the visualization is simple. Consider Figure A.1, which places a voter at the origin and an incumbent at $(1, 0)$.

In Figure A.1, if the challenger is located anywhere in either Region A or Region B, then she will be less "extreme" in the first dimension than the incumbent, but the voter located at the origin will still prefer the incumbent because the incumbent is less "extreme" in the second dimension.

Figure A.1 shows how a single voter, located at the origin, can prefer an "extreme" incumbent to a more "moderate" challenger because of an additional policy dimension in which the challenger is less moderate. However, this figure only shows what happens with a single voter. With multiple voters in a multidimensional policy space, the basic problem is that there can be no "Condorcet winner," which defeats any other location in a one-on-one contest. Rather, for any location, no matter how "moderate" it looks, there will be a second location that can defeat it in a one-on-one contest. So we cannot simply find some geographic center in a cloud of voters' ideal points, such as the origin in Figure A.1,

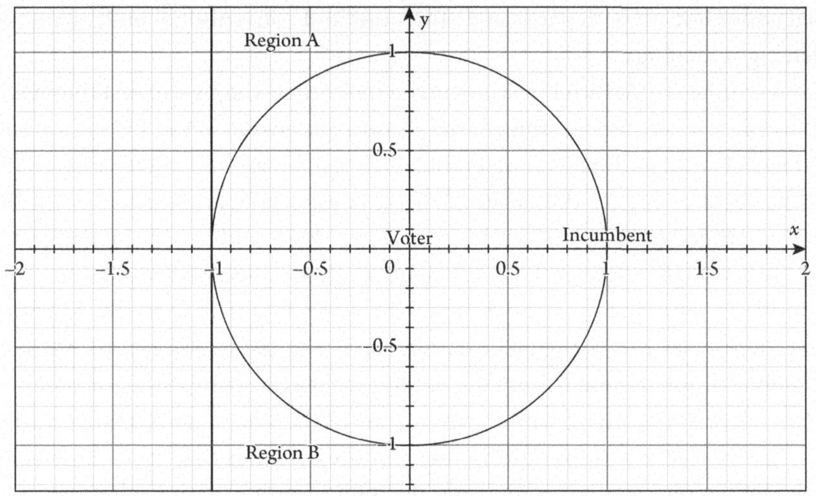

Figure A.1 Preference for an "extremist" in two dimensions

call it centrism, and wonder why candidates win at other locations. To do so misses the point of multidimensionality. In a multidimensional policy space, an incumbent located at $(r, 0, 0, \ldots, 0)$ can win, but in a multidimensional policy space, a candidate located *anywhere* can win, given the right sequence of elections (McKelvey 1976).

The question becomes how compelling an explanation this is for the electoral success of extremely liberal or extremely conservative incumbents. In fact, that is rather difficult to assess. While we have methods for estimating challengers' liberalism or conservatism based on policy statements, interest group contributions, and other factors, assessing their locations in the policy dimensions whose substantive meanings are less clear would be deeply problematic. While Poole and Rosenthal's procedure allows us to uncover more than simply the first dimension in roll call voting—ostensibly liberalism-conservatism—the fact that these other dimensions have meanings that are unclear today makes it especially problematic to try to assess where challengers would be, even if we can estimate their locations along something like the first dimension through other means.

For the mid-20th century, that would be less of a problem, when the second dimension of NOMINATE reflected a regional and racial aspect to political divisions. However, as Noel (2014) has argued, liberalism and conservatism are redefined over time by coalition merchants and policy demanders. Over the course of the second half of the 20th century, issues of race were folded into liberalism and conservatism, producing the modern era in which the first dimension of NOMINATE explains far more about roll call voting, and in which racial issues line up with partisan distinctions and conventional understandings of left-right ideology.

Thus, while NOMINATE can estimate more than simply the first dimension of the policy space, the meaning of the second dimension in the 113th Congress is far less clear than the meaning in the 83rd, to scholars and voters alike (much less the third or higher dimensions). So it is especially unclear how we could measure where challengers are located along these additional dimensions or how voters might place sufficient weight on these dimensions so as to prefer incumbents who seem extreme along the liberal-conservative dimension to spare themselves the challengers' extremism along the less substantively interpretable additional dimensions that we would have difficulty estimating anyway.

Not surprisingly, then, while "extremism" of the liberal-conservative variety has negative electoral implications, there is little evidence that other dimensions of NOMINATE have any significant electoral impact in the modern era, now that liberalism and conservatism have been redefined.

Next, consider the strategic logic of incumbents winning on the basis of other policy dimensions. Once we enter the multidimensional policy space, collective preferences become intransitive, and there are no Condorcet winners. Therefore, incumbents' locations cannot be Condorcet winners. Therefore, a hypothetical challenger *could* defeat *any* incumbent with the right positioning decision.

The result would be a paradoxical and nonexistent *challenger*'s advantage. Since incumbents adopt locations through their roll call voting patterns, by the time of an election, they are locked into positions that cannot be Condorcet winners because no such positions exist. Challengers, unconstrained because they have no voting records, can promise a location that defeats the incumbent's position, and such a counterposition necessarily exists because the policy space is multidimensional. Rather than multidimensionality *explaining* incumbents' victories, it would put incumbents at a structural *dis*advantage with regard to positioning decisions. Rather than overwhelming incumbent success, we would see overwhelming incumbent loss. After all, since nonpivotal legislators can costlessly position themselves anywhere, by the reasoning in chapter 3, challengers wouldn't even have a credibility problem.[1] Congressional terms would follow the one-and-done rule. Clearly, that isn't the case.

Multidimensionality becomes an even less compelling explanation for the success of extreme incumbents in an era of homogeneous parties. In principle, it is possible for an incumbent with an ostensibly extreme position to defeat a challenger who looks more moderate by exploiting dimensions other than liberalism-conservatism, but that explanation does not work if incumbents simply aren't taking those positions.

Consider how we might operationalize party ideological homogeneity in a standard multidimensional context, representing the circumstances in which the conditions for party government are met (and as we shall see later, those conditions are identical in a multidimensional model anyway). Suppose that all Republican candidates are constrained to fall within $\varepsilon > 0$ in Euclidean distance from $(R_1, R_2, \ldots, R_n)$, and all Democrats are constrained to fall within $\varepsilon > 0$ in Euclidean distance from $(D_1, D_2, \ldots, D_n)$. If ε is sufficiently small, reflecting party homogeneity, then even if there is some platform with which the incumbent could defeat a "moderate" challenger despite that platform being "extreme" along the liberal-conservative dimension, if that winning platform is not within ε of her *party* platform, its electoral benefits are irrelevant because nobody within the homogeneous party is adopting such platforms. Thus, party homogeneity—a condition for party government—will mean candidates won't do what is necessary to use multidimensionality to defeat more "centrist" opponents.

The policy space is multidimensional. However, that does not mean all models must be multidimensional. If our goal is to explain intraparty conflict among congressional Democrats in the mid-20th century, we clearly cannot do that with a unidimensional model. However, the goal here is to explain the electoral success of ideological extremists in the late 20th and early 21st centuries. If such incumbents are not winning *because of* the multidimensional nature of the policy space, then we can exclude such dimensions for the sake of parsimony and tractability. As we shall see next, though, the underlying principles of the model can be adapted to multidimensional policy spaces anyway, with only somewhat different implications.

Multidimensionality and the Incremental Positioning Model

Consider the electoral component of the model. Multidimensional spatial models can generate unusual election results, but conventional methods of constructing them still assume proximity voting at the individual level. Once we consider the incremental nature of the positioning process, even that breaks down, just as in the unidimensional case.

Chapter 2 demonstrated that two types of phenomena can produce a proximity paradox for an individual voter: asymmetric frequencies and asymmetric differentials. Both are equally applicable regardless of dimensionality. Asymmetric frequencies occur whenever there are regions in the policy space with relatively few cut points for roll call votes. Clearly, the same phenomenon can occur in an *n*-dimensional policy space. In the crudest example, suppose that a voter's location in an *n*-dimensional space is $V = (v_1, v_2, \ldots, v_n)$, the incumbent legislator's location is $L = (l_1, l_2, \ldots, l_n)$, and the challenger's location is $C = (c_1, c_2, \ldots, c_n)$. Suppose that the voter has a quadratic loss utility function, and that the Euclidean distance between V and L is greater than the Euclidean distance between V and C.

Any two points are collinear. Consider the line running through V and L. If C does not fall along that line, then there must be an infinite number of hyperplanes parallel to the line created by V and L, which separate V and L from C. For any vote on which the status quo and alternative divide the policy space along one of those hyperplanes, the incumbent's vote would be consistent with the voter's preferences, and the challenger's hypothetical vote would not. Let S_1 be the set of all such roll call votes. Similarly, let S_2 be the set of all roll call votes that separate V and C from L with hyperplanes parallel to the line created by V and C. Suppose, further, that all agenda items are elements of either S_1 or S_2. If S_2 is relatively empty, then asymmetric frequencies will lead the voter to favor the incumbent despite the voter's proximity to the challenger, just as in the unidimensional case.

Asymmetric differentials in the *n*-dimensional policy space are no more complex. Since the voter's utility function is continuously defined, as the status quo and alternative points for any given roll call vote approach each other, the voter's utility differential between the two approaches zero asymptotically. Thus, if the status quo and alternative points in S_2 are arbitrarily close together and those in S_1 are not, then the voter will prefer the incumbent's bundle of roll call votes even when the sets have the same number of elements.

Neither asymmetric frequencies nor asymmetric differentials are unique to unidimensional policy spaces, then. It is their implications that differ in the multidimensional context. In an era of homogeneous parties, when Democrats are within ε of $(d_1, d_2, \ldots, d_n)$ and Republicans are within ε of $(r_1, r_2, \ldots, r_n)$, all candidates are clustered around one of two points. Since any two points are collinear, the policy space is close to unidimensional because nobody adopts positions far enough away from their party's platforms to exploit the chaos theorem. But what about when the parties are *not* ideologically homogeneous? In the mid-20th century, far less about congressional roll call voting could be explained by a simple, liberal-conservative dimension because of the complex interaction between race, region, and party. McKelvey shows that these situations are infinitely exploitable, but the observations here suggest that the problem is even worse. Unmoored by proximity voting because of the potential impact of asymmetric frequencies and asymmetric differentials, nearly any election result becomes explainable to a degree that even McKelvey didn't argue. Thus, in an era of heterogeneous parties, elections are potentially even more "chaotic" than McKelvey would suggest.

Finally, though, additional consideration should be paid to the unified model in chapter 4 and the conditions for party government. Chapter 4 reduced a legislator's willingness to vote for party discipline to three considerations: the policy costs of voting for party discipline, the policy benefits of doing so, and the electoral costs of acceding to discipline. The inequalities that govern this vote are, in fact, identical regardless of dimensionality, so the conditions for party government remain ideological homogeneity, electoral heterogeneity, and strong policy preferences regardless of the dimensionality of the policy space. Even if the legislature is not purely unidimensional, the logic of weighing policy and electoral considerations remains the same. However, recall that party homogeneity implies an effectively unidimensional policy space anyway since two homogeneous parties are clustered around two points, and any two points are collinear.

Appendix B

IDEOLOGICALLY UNSOPHISTICATED VOTERS AND BUNDLES OF ROLL CALL VOTES

Chapter 2 developed a model of voting behavior in which ideologically sophisticated voters respond, retrospectively, to the bundles of roll call votes that incumbent members of Congress cast in order to establish their locations in the policy space. The model suggested that ideologically extreme incumbents can defeat more centrist challengers because voters with single-peaked, symmetric preferences over policy locations can have paradoxical preferences over the bundles of roll call votes associated with those locations. The model was built around hypersophisticated voters in order to show that voting for the more distant candidate is not a mistake resulting from imperfect information, bias, calculation error, or any other flaw in their reasoning. By stacking the deck in favor of proximity voting through the assumption of hypersophisticated voters and generating violations of such preferences, the model shows how weak normal Downsian pressure is with respect to legislative elections. Nevertheless, one might wonder how results would change if voters are less sophisticated. Consider a model of voting behavior built more closely around Converse (1964). Converse argued that most voters have little understanding of the traditional liberal-conservative spectrum around which spatial models are built. Rather, Converse argued that the public can be divided into issue publics, who each care about a small number of issues. This appendix builds a spatial model of a legislative election combining bundles of roll call votes with an issue public-based electorate, which is less cognitively sophisticated than the electorate in the model from chapter 2. The predictions are substantively identical to those derived from the hypersophisticated electorate in chapter 2. Incumbents receive electoral rewards for movement toward the center, but can still defeat more centrist opponents.

An Ideologically Unsophisticated Electorate

There will be an incumbent and a challenger. The incumbent's location in the policy space will initially be fixed and exogenous. For reasons to be explained shortly, the challenger's location will not be a parameter in the model, as in the unified model from chapter 4. The electorate will be divided into three categories: partisan voters, issue voters, and stochastic voters.

1. **Partisan voters.** According to dominant models of voting behavior from political psychology (those following from Campbell et al. 1960), the most important determinant of vote choice is party identification. Spatial models generally exclude partisanship entirely, and most that include it generally follow from Erikson and Romero (1990), by including partisanship as merely another dimension, distinct from policy,[1] within which voters have preferences. Somewhat differently, Sniderman and Stiglitz (2012) treat partisanship as the result of a spatial preference for a party's policy program, which complicates the choice between candidates' positions when those candidates deviate from their party's programs. Other models, such as Adams et al. (2005), have considered the possibility that partisanship induces spatial bias, following from which Jessee (2012) proposes a psychology-based model built on the dynamic between partisanship and ideological perception. So partisanship has not gone entirely ignored by spatial theorists. However, none of these models of partisanship reflect the conceptualization of party identification as group identity. Converse (1964) argued that a plurality of voters fall into a "level of conceptualization" defined by "group interest." Following from that, Green et al. (2002) argue that partisanship is mostly a reflection of associations between group membership and party coalitions. In that school of thought, partisanship does not merely condition ideology—it simply determines voting behavior for the lion's share of the electorate. In order to build a model in the Converse tradition rather than the Downs tradition, then, this appendix will assume that partisan voters simply have their votes determined by party rather than policy locations.[2] The partisan balance of the district serves as the baseline from which the incumbent incrementally affects her vote share through her roll call voting. There is a fixed number of incumbent partisan voters in the district, given by V_{MC}, and a fixed number of challenger partisan voters in the district, given by V_C.
2. **Issue voters.** Issue voters are divided into issue publics. As in chapter 2, we shall assume a single policy dimension. A policy relating to *any* issue can be characterized by a location in that policy space. So a specific abortion policy

can have a location of p^*, indicating a specific degree of liberalism. A tax policy with a location of p^*, then, would have the same degree of liberalism as the aforementioned abortion policy. For any given policy, p, a member of issue public j will have the following utility function, where $\mathbf{1}(x)$ is an indicator function taking a value of 1 if x is true, 0 if x is false:

$$U_{issue\ j voter}(p) = \mathbf{1}(p \text{ involves issue } j)\left[-\left(p - i_{voter}\right)^2\right] \quad \text{(B.1)}$$

Thus, an issue voter is indifferent between policy alternatives unless those alternatives involve the one issue with which the voter is concerned, making the voter less ideologically sophisticated than the voters in chapter 2, and more similar to voters in Converse's (1964) description.

Issue public ideal points for issue j in a district are normally distributed with a mean of θ_j and a variance of σ_j^2. Let us assume that there are N issue publics, whose total populations are given by the following vector: $[S_1\ S_2 \ldots S_j \ldots S_N]$.

3. **Stochastic voters.** Stochastic voters' preferences are unclear, but we will estimate that, on average, they will behave like the rest of the voters in the district. At any given time, t, the proportion of stochastic voters expected to support the incumbent's reelection will be a draw from a normal distribution with mean μ_t and variance σ_t^2, where μ_t is the estimate at time t of the proportion of all *other* voters who will vote for the incumbent in the upcoming election. Let us assume there are R stochastic voters in the district. Elections are always somewhat uncertain.

Behavior of Issue Voters

Partisan voters' decisions are fixed by their party identification, and stochastic voters will behave, on average, like the rest of the district, so an election will hinge on how issue voters respond to roll call votes. Since issue voters have no ideologies, we shall assume that issue voters do not observe candidates' ideologies. After all, according to Converse (1964), voters have little comprehension of liberalism and conservatism anyway. Instead of observing candidates' locations, issue voters observe only roll call votes. Since a member of issue public j has no partisan preferences and no policy preferences on any issues other than issue j, an issue voter will be indifferent between any incumbent and any challenger unless the legislative agenda includes votes that address the one issue about which that voter cares. Thus a member of issue public *j* will abstain from the election unless there is a roll call vote related to issue *j*.

Each roll call vote is a choice between a liberal option, *l*, and a conservative option, *c*. For the sake of simplicity, let us assume that each roll call vote addresses an issue of concern to precisely one issue public, and that no two roll call votes address the same issue. This assumption will allow us to predict the effect of any given incumbent roll call vote on that incumbent's probability of re-election without considering the positions of the challenger, as we shall see. Given this restriction, any time an incumbent votes in a manner consistent with an issue voter's preference, that issue voter will vote for the incumbent, and any time an incumbent votes in a manner inconsistent with the preferences of an issue voter, that issue voter will vote for the challenger.

This decision rule follows from the one-vote-per-issue rule. If single-issue voters have no comprehension of ideology, then they cannot impute hypothetical roll call votes to the challenger given the challenger's ideology, in which case they have nothing to which to compare the incumbent's voting record. Furthermore, challengers have strategic incentives to blur their policy positions (see, for example, Page 1978; Shepsle 1972; Tomz & Van Houweling 2009).

Talk is cheap. If issue voters lack the cognitive sophistication to observe candidates' locations in the policy space, then the only information they have with which to make their decisions is the incumbent's roll call voting record. Moreover, if no issue receives more than a single roll call vote, the members of any one activated issue public only have a single roll call vote on which to base their decisions. If Voter A agrees with the incumbent's vote, then she cannot be certain that she will agree with future votes that the incumbent would cast if re-elected, but she has more reason to believe that than with a challenger about whom she has no information. Thus, Voter A should vote for the incumbent. On the other hand, if Voter B disagrees with the incumbent's roll call vote, she cannot be certain that she will disagree with future votes that the incumbent might cast, nor that she would agree with the votes that the challenger would cast. However, given her disagreement with the incumbent, she has little to lose and much to gain by voting for the challenger. So Voter B should vote for the challenger. Thus, if we assume one roll call vote per issue, we can model the behavior of ideologically unsophisticated issue publics without considering the challenger's location, which unsophisticated voters cannot evaluate anyway. Thus, we will assume that each roll call vote activates a new issue public, that every issue voter who agrees with the incumbent's vote will vote for the incumbent, and that every issue voter who disagrees with the incumbent's vote will vote for the challenger.

Responses to Roll Call Voting Records by an Ideologically Unsophisticated Electorate

With the actors defined as above, we can model an ideologically unsophisticated electorate's response to the incremental process by which legislators adopt their locations in the policy space. By assumption, the legislative agenda consists of N roll call votes, with each roll call vote involving one unique policy area, thereby activating a unique issue public. For any given roll call vote, *j*, there is a cut point located at $\left(c_j + l_j\right)/2$, such that any actor with preferences over that issue with an ideal point located at $i < (c_j + l_j) / 2$ prefers the liberal option and any actor with preferences over that issue with an ideal point located at $i > (c_j + l_j) / 2$ prefers the conservative option. If the incumbent casts a sequence of roll call votes in order to incrementally adopt a location of MC, how will the legislative session and subsequent election proceed?

The voting record that the incumbent will build depends on the relationship between the incumbent's spatial location, MC, and the vector of cut points, $[(c_1 + l_1) / 2 \ (c_2 + l_2) / 2 \ \ldots \ (c_N + l_N) / 2]$. For any roll call vote, *j*, if $MC < (c_j + l_j) / 2$, the incumbent votes for the liberal option, and if $MC > (c_j + l_j) / 2$, she votes for the conservative option. The purpose of the model is to examine the electoral effects of an ideologically extreme voting record, so let us begin by examining what would happen if $MC > (c_j + l_j) / 2$ for all *j*. Thus, the member of Congress will vote for the conservative option on every bill. For any given issue public *j*, then, any voter with an ideal point to the left of $(c_j + l_j) / 2$ will vote for the challenger, and any voter with an ideal point to the right of $(c_j + l_j) / 2$ will vote for the incumbent. Let F be the cumulative distribution function for issue voters' ideal points within issue public *j*, a normal distribution. For any specific roll call vote *j*, then, the number of issue voters who are mobilized in favor of the incumbent will be $S_j[1 - F([(c_j + l_j) / 2 - \theta_j] / \sigma_j)]$, and the number of issue voters who are mobilized against the incumbent will be $S_j[F([(c_j + l_j) / 2 - \theta_j] / \sigma_j)]$. Thus, the combined totals of partisan voters and issue voters voting for and against the incumbent are given by the following formulae:

$$\text{nonstochastic incumbent votes} = V_{MC} + \sum_{j=1}^{N} S_j \left[1 - F\left(\frac{\frac{c_j + l_j}{2} - \theta_j}{\sigma_j} \right) \right] \tag{B.2}$$

$$\text{nonstochastic challenger votes} = V_{C} + \sum_{j=1}^{N} S_j \left[F\left(\frac{\frac{c_j + l_j}{2} - \theta_j}{\sigma_j} \right) \right] \tag{B.3}$$

Thus, the incumbent's expected vote share given her legislative voting record will be given by the following:

$$\text{incumbent's expected vote share} = \frac{V_{MC} + \sum_{j=1}^{N} S_j \left[1 - F\left(\frac{\frac{c_j + l_j}{2} - \theta_j}{\sigma_j} \right) \right]}{V_{MC} + \sum_{j=1}^{N} S_j \left[1 - F\left(\frac{\frac{c_j + l_j}{2} - \theta_j}{\sigma_j} \right) \right] + V_C + \sum_{j=1}^{N} S_j \left[F\left(\frac{\frac{c_j + l_j}{2} - \theta_j}{\sigma_j} \right) \right]} \tag{B.4}$$

This is an expected value, of course, and stochastic voters will add an element of random error to the actual election result, but the formula above describes the central tendency of the electoral consequences of the incumbent's decision to vote for the conservative option on every bill, thereby incrementally adopting an ideologically extreme location in the policy space.

What does this mean for the electoral response to an extreme voting record? Chapter 2 showed that an electorate consisting of hypersophisticated ideologues would respond to legislative voting records by sometimes electing an extremist over a centrist, but providing increasing vote shares to candidates who move toward the location of the median voter. How different is this from the predictions of Equation B.4 above? Hardly at all.

Let us consider two possible cases. In Case 1, we will assume that the incumbent's ideological position is more conservative than the median voter of each issue public. This will represent the circumstance of an incumbent who is, in some sense, too ideologically extreme for the district (or at least the analog to that when no individual voter actually has a complete ideology). In Case 2, we will assume that each issue public median voter has an ideal point translating to the same location in the complete policy dimension, which is identical to the incumbent's location in the policy space. This will represent the circumstance of an incumbent who incrementally adopts what is functionally the location of the district median voter.

First, consider Case 1. Suppose that $MC > \max\{[\theta_1, \theta_2, \ldots, \theta_N]\}$. Can such an extremist win re-election with an expected vote share given by Equation B.4?

There are several simple ways that the incumbent can win. The first, and most obvious, is if $V_{MC} > V_C$ (meaning that the district contains more incumbent partisans than challenger partisans). In general, we should expect the district partisan balance to favor the incumbent. If the incumbent begins the legislative session with an electoral advantage, then even if every individual roll call vote adds more votes to her challenger's expected total than to her own, as long as the gap created by her voting record is less than the incumbent's initial partisan advantage, she will end the session with an expected vote share above 50%. Of course, this is an uninteresting possibility because if this is the explanation for the electoral success of extremists, then we are explaining deviations from spatial theory with nonspatial voting. While this may be a substantively important process, it does more to undermine spatial models of elections than to advance them.

A similarly uninteresting possibility of incumbent victory is created by the stochastic voters. Even if the incumbent's ideologically extreme voting record gives her an expected vote share below 50%, stochastic voters might hand her a re-election victory due essentially to voter error. Obviously, this too is an uninteresting possibility from a spatial theory perspective.

So let us turn to the legislative agenda. Are there legislative agendas that favor the incumbent's re-election given that $MC > \max\{[\theta_1, \theta_2, \ldots, \theta_N]\}$. Yes. The most simple example would be if there are no roll call votes, *j*, such that $MC > (c_j + l_j) / 2 > \min\{[\theta_1, \theta_2, \ldots, \theta_N]\}$. If so, then the incumbent member of Congress will never cast a roll call vote against the preferences of the majority of the relevant issue public in her district. Of course, we rejected such arguments in chapter 2 in part because if this were the case, then incumbent locations would never be observationally different from district preferences.

In order to show a meaningful victory by an ideologically extreme incumbent, we must understand one crucial observation about the arithmetic of this model. Each roll call vote, *j*, such that $MC > (c_j + l_j) / 2 > \theta_j$ will add more raw votes to the challenger's nonstochastic vote total than to the incumbent's nonstochastic vote total, thereby reducing the raw vote margin of the incumbent. However, any roll call vote, *k*, such that either $(c_k + l_k) / 2 > MC > \theta_k$ or $MC > \theta_k > (c_k + l_k) / 2$ will add more raw votes to the *incumbent's* nonstochastic vote total than to the challenger's vote total, thereby *increasing* the incumbent's raw vote margin. In order for the legislative agenda to allow an extreme incumbent to win re-election while still allowing us to observe her extremism, all that is necessary is for the legislative agenda to include a sufficient number of roll call votes that activate issue publics in favor of the incumbent to compensate for the roll call votes that activate other issue publics against her.

So consider an incumbent who begins the legislative session with a 60% to 40% partisan advantage in a district with 200,000 partisan voters. Before any votes are cast, the incumbent expects to have 120,000 votes in her favor and 80,000 votes against. Suppose that the incumbent's location in the policy space is 3. The legislative agenda consists of two bills. Bill 1 moves abortion policy from 0 to 1. Bill 2 moves tax policy from −3 to −1. Finally, suppose that the issue public for abortion and the issue public for tax policy consist of 10,000 voters each where $\theta_{abortion} = \theta_{tax} = 0$, and $\sigma_{abortion} = \sigma_{tax} = 1$. We can now apply Equation B.4.

The incumbent wins the votes of 120,000 partisan voters automatically, and the challenger wins the votes of 80,000 partisan voters automatically. Bill 1 is associated with a cut point of 0.5, which is 0.5 standard deviations to the right of the median voter's ideal point in the abortion issue public. With 10,000 abortion voters, the consequence of the incumbent's conservative vote on abortion is to mobilize 6,915 voters for the challenger, and 3,085 voters for herself. However, Bill 2 is associated with a cut point of −2 on tax policy. With 10,000 tax voters, the consequence of the incumbent's conservative vote on tax policy is to mobilize 228 votes for the challenger and 9,772 votes for the incumbent. This yields the following vote totals from nonstochastic voters:

$$\text{nonstochastic incumbent votes} = 120{,}000 + 9{,}772 + 3{,}085 = 132{,}857 \tag{B.5}$$

$$\text{nonstochastic challenger votes} = 80{,}000 + 6{,}915 + 228 = 87{,}143 \tag{B.6}$$

With those nonstochastic vote totals, the incumbent's expected vote share *rises* from 60% before the legislative session to 60.4% after the session. Moreover, her raw vote margin over the challenger rises from 40,000 to 45,714. Thus, in order for the challenger to win, she must now win among stochastic voters by a margin greater than 45,714 rather than a margin greater than 40,000 with the same, fixed number of stochastic voters (and a slightly lower *expected* vote share among them). Despite the incumbent's position to the right of the median in each issue public, and despite a voting record that allows sophisticated observers to measure that location and observe the deviations from voter preferences, the incumbent's voting record increases her probability of victory relative to her initial baseline.

At this point, it would be tempting to infer that extremism can be paradoxically beneficial in the issue public model. However, such an inference would be based on comparing the incumbent's vote total given the hypothetical legislative agenda above to the wrong baseline. Rather than comparing her expected vote share after the legislative session to her initial baseline, we must compare her expected vote share after the legislative session to what her expected vote share would have been had she incrementally built a more moderate voting record. That brings us to Case 2. So now suppose that $MC^* = 0$. Given the same legislative agenda and same electorate, how would a legislator located at MC^* fare?

A location of 0 implies voting for the liberal option on Bill 1 and the conservative option on Bill 2, yielding the following vote totals from nonstochastic voters:

$$\text{nonstochastic incumbent votes} = 120{,}000 + 9{,}772 + 6{,}085 = 136{,}687 \tag{B.7}$$

$$\text{nonstochastic challenger votes} = 80{,}000 + 3{,}085 + 228 = 83{,}313 \tag{B.8}$$

The incumbent would receive 136,687 votes from nonstochastic voters, with her challenger receiving 83,313, for an expected incumbent vote share of 62.1%. This is a critical baseline to which we should compare our extremist legislator's vote share. With an ideologically extreme voting record, a legislator located at 3 would expect to receive 60.4%, which is *higher* than her initial expectation despite her ideologically extreme voting record. However, had she incrementally built a more moderate voting record, she would expect to win an even higher vote share of 62.1%. Thus, precisely as in chapter 2 with sophisticated voters, an extremist can win, but there are more rewards for centrism.

Generalized Election Results with Unsophisticated Voters

The simple demonstration above shows us that an ideologically extreme incumbent can paradoxically increase her vote share above her initial baseline through the roll call voting process. That will allow ideologically extreme legislators to win re-election, even when their extremism is observable. Since voters in this model are not ideologically sophisticated, we should not be surprised that ideological extremists can win, although their ability to gain from the roll call voting process may not be expected. What is more interesting is the comparison between the extreme and moderate voting records. For any given roll call vote that activates a unique issue public, the incumbent faces a choice between adding X voters to her own nonstochastic vote total and Y to her opponent's total, or adding Y to her own nonstochastic vote total and X to her opponent's. If X > Y, the electorally optimal choice must always be the former. If we assume that $\theta_j = \theta_k$ for all issues, *j* and *k*, so that there is a location analogous to a district median voter, then the electorally optimal location for a legislator to adopt incrementally must be θ_j. This is because incrementally adopting any other location means casting at least one roll call vote that adds more nonstochastic votes to the challenger's total than to her own. Arithmetically, that cannot be electorally optimal. A legislator's decision *not* to incrementally adopt a location of θ_j does not mean she will lose, but she will receive a lower vote share than an incumbent from a similar district who *does* incrementally adopt a location of θ_j.

So what does this tell us about how an electorate composed of ideologically unsophisticated issue publics will respond to a legislative voting record? An incumbent who builds an ideologically extreme voting record can win, not because voters deviate from spatial reasoning, but because a legislative agenda can mobilize more issue voters in favor of the incumbent than against the incumbent, even when the incumbent is an extremist compared to the preferences of the various issue publics in her district. However, if a legislator in the same district were to build a more centrist voting record, that legislator would win a higher vote share than the extremist would have won. Thus, the unsophisticated electorate segmented into issue publics will behave, in aggregate, much like the ideologically sophisticated electorate from chapter 2. It will apply modest electoral punishment for ideological extremism, but the modesty of that punishment will allow many extreme incumbents to win re-election anyway.

One of the most common critiques of spatial theory from political psychologists is that voters cannot perform the tasks that we assume, but the model here shows that the critique is less important, in aggregate, than political psychologists suggest. The model in this appendix makes minimal assumptions about the sophistication of voters, who know virtually nothing and make simple-minded decisions. In aggregate, though, they behave the same way as the hypersophisticated electorate in the model from chapter 2. They reward movement toward the center with higher vote shares, but frequently allow extremist incumbents to win re-election. If we are concerned with aggregate rather than individual behavior, then, the cognitive sophistication of voters is less important than political psychologists suggest. Regardless of their individual capabilities, the electorate will collectively provide merely a small but nondeterminative advantage to centrists, frequently allowing extremist incumbents to win re-election, all depending on the legislative agenda.

NOTES

Chapter 1

1. In a multidimensional context, of course, there is no median voter, which doesn't just permit other outcomes in equilibrium—it frequently precludes pure strategy equilibria. While this book will rely primarily on unidimensional models, appendix A will address extrapolations to and implications of multidimensionality.
2. Groseclose and Milyo (2010) derive this basic result in the context of legislators' decisions about how to vote on poison pill amendments. The underlying logic is the same here, though.
3. Throughout the book, a "sincere" roll call vote will mean one that reflects a legislator's personal policy preference, and an "insincere" vote will be one that conflicts with the legislator's policy preference.
4. There are variations which include a postelection bargaining stage, such as the organization of the legislature or negotiation between the executive and the legislature. These variations will be discussed later. However, even in these variations, the final determination of policy is a single action represented by one point in the policy space.

Chapter 2

1. The discharge petition is an unusual maneuver in which 218 signatories can force the Speaker to bring a bill that has stalled in committee to the floor under a closed rule. However, reaching 218 requires convincing majority party members to sign the petition, and party leaders look harshly on individuals who undercut the majority party's procedural control. Thus, discharge petitions are rarely successful because even when majority party members want to pass a bill, they are generally unwilling to face the punishment that may be incurred by signing a discharge petition.
2. In the original derivation of the proximity paradox (Buchler 2011a), this was the basis of the argument that the phenomenon benefits majority party legislators specifically.
3. Previous demonstrations constrained the initial status quo points across issue areas to be identical to show that a proximity paradox can occur even if we make the highly restrictive assumption of uniform initial status quo points. Of course, it would be empirically absurd to argue that all status quo points for all policy areas are equally liberal or conservative at any given point in time. With a few notable increases, the general trend in tax rates since the early post-World War II peak has been a decrease, but since the Great Depression, we have seen a periodic but dramatic expansion of federal entitlements, from Social Security to Medicare to the healthcare subsidies in the Patient Protection and Affordable Care Act. Either tax rates and entitlements were not equally liberal before these trends, or they are not equally liberal now (with this going to press at a time when a partial "repeal-and-replace" of the ACA appears to have failed in the Senate after the Republicans' capture of the House, Senate and Presidency in

2016). However, we cannot reconcile these long-term policy trends with spatially consistent status quo points for all policy issues across time. So it was mathematically interesting, but empirically unimportant, that previous results could be generated with uniform initial status quo points. Nevertheless, it is worth noting that the following possibility does require a wide distribution of initial status quo points because it will require voters to be relatively indifferent between status quo and alternative points for roll call votes with cut points far from their ideal points in both directions.

4. Of course, this effect will not help liberal-leaning moderates in conservative-leaning districts. Quite the opposite, which helps to explain the negative electoral impact of voting for the PPACA in 2010 among Democrats representing at least marginally Republican districts.

Chapter 3

1. However, Lott and Reed (1989) have argued that if legislators position themselves sincerely, the electoral process will sort out the legislators whose positions are inconsistent with their constituents' preferences, so over time, legislators' sincere preferences and their constituents' preferences will converge. This argument has been posed as an explanation for the lack of a notable change in retiring legislators' voting habits (Bender & Lott 1996; Lott 1987; Lott & Davis 1992).
2. In principle, competing pressures might pull legislators in either direction, either toward or away from the center. This question will be addressed in subsequent chapters. For example, primary pressure may pull candidates in the opposite direction from general election pressure. For the purposes of this chapter, though, we will focus on unidirectional pressure.
3. DeBacker (2012) has argued that parties can constrain legislators by creating incentives related to their postlegislative careers. If so, then retiring legislators might still be subject to party pressure even after their retirement decisions. If so, then perhaps their postretirement decision voting behavior is not sincere. Parker (2008), though, argues that incumbents have a variety of ways to set up their own postretirement careers.
4. In chapter 6, we will discuss the House vote on the PPACA, and public statements by Representative Bart Stupak (D-MI) make this a somewhat more complicated determination for him.
5. Of course, Lieberman's problem with primaries was relevant here too.

Chapter 4

1. There is a caveat. The results of Groseclose and Milyo's (2010) model are actually indeterminate if everyone doing so produces a margin of victory or defeat for the bill of a single vote. In that instance, there are no pure strategy equilibria. However, this is a minor issue because so few roll call votes are decided by a single-vote margin, which will be central to the argument here.
2. In essence, then, what this chapter will call an "insincere" vote is analogous to what Groseclose and Milyo called a "sincere" vote. The difference in terminology reflects both the fact that this chapter will not assume, as Groseclose and Milyo did, that a legislator and her constituents have the same policy preferences, and that this chapter will focus on final passage votes in a unidimensional model rather than amendment votes in the face of potentially cyclical preferences within the legislature.
3. In a conventional model of negative agenda control, for example, the agenda-setter will not permit any floor votes with cut points between her own ideal point and the floor median if the alternative is further from her ideal point than the status quo.
4. Note that either MC > PL or MC < PL could be compatible with this circumstance, so whether the member of Congress is more or less extreme than the leader is irrelevant here.
5. It is important to understand, then, that "electoral heterogeneity" has a very different meaning here than in Harden and Carsey (2012). Harden and Carsey argue that "electoral heterogeneity" strengthens parties by pulling legislators away from their median voters, which sounds very much like the argument here. However, Harden and Carsey use the term "heterogeneity" to refer to the spread of constituent opinion within a state around the state's median voter

(they examine the Senate rather than the House). Here, though, "heterogeneity" refers to the spread of district median voters within a legislative caucus. Harden and Carsey use similar terminology to describe a very different process.

6. In principle, this condition could be met with an unusual form of electoral homogeneity. Suppose that every caucus member represents a district with equally centrist preferences. If so, then the caucus is electorally homogeneous, but there will still be many elements in the first series on the right-hand side of Inequality 4.6. However, despite the historical fantasies of redistricting reform advocates, this has never been empirically true (and chapter 5 will demonstrate that there hasn't even really been a decline in the proportion of legislators who represent centrist districts). Every party caucus includes a large contingent of members who represent heavily slanted districts, so realistically, condition (b) implies electoral heterogeneity.
7. It should be noted that there is an indirect relationship between this condition and Patty's (2008) model. Patty argued that a narrow majority creates more incentives for a caucus to delegate power to party leaders. In the context of the model here, the more narrow the majority is, the less electoral heterogeneity is necessary to make Inequality 4.6 true. A 220-member caucus only needs three cross-pressured members to defect in order to lose a roll call vote, even when the remaining 217 members have similar constituencies. On the other hand, a 250-member caucus can accommodate 32 defections without losing a roll call vote, so a larger caucus must be more heterogeneous in order to create the conditions for party government.
8. Note that this argument conflicts with some conceptualizations of CPG, although not all. Ladewig (2005) makes a distinction between preferences and interests and argues that homogeneity of constituent *interest* is a condition for party government. Since this model focuses on preferences rather than interests, these arguments are not entirely incompatible.
9. DeBacker (2012) argues that retiring legislators still have incentives to obey party leaders because party leaders can help them to secure future positions. However, it is difficult to reconcile that with the behavior of "true" retirees—those who no longer harbor electoral ambitions and either want to leave political life, or take jobs as lobbyists that they have secured for themselves, which is within the capacity of legislators to arrange for themselves (Parker 2008).

Chapter 5

1. Chapter 7 will address the divided government period from 2011 through 2016 separately based on the observation that party leadership adapted in the 112th Congress to the unique problems created by combining divided government with extreme polarization.
2. A Congress directly following a presidential election will be an odd-numbered Congress, and given the potential for measurement error and preference change over time, it is best to use Congresses immediately following the time at which district preferences are measured.
3. In fact, it is more complicated than that, as we shall see with Senator Lisa Murkowski (R-AK) and former Senator Joe Lieberman (D-CT).
4. Of course, that favor diminished when Rubio began advocating immigration reforms that were unpopular among the Republican activist base, and by the time of his 2016 presidential campaign, he was considered to be an "establishment" figure in the party.
5. Shira Toeplitz, "John Sullivan's Big Mistake? Ignoring Challenger for Too Long," *Roll Call*, June 28, 2012.
6. As reported by Steve Kornacki in "The Price of Tribal Betrayal," *Salon*, June 1, 2012.
7. Jenkins and Nokken (2008) argue that during a lame-duck session, exiting members cease to be subject to either party or electoral pressure, but returning members are subject to a combination of reduced electoral pressure but equally strong party pressure. However, if electoral pressure varies by proximity to the next election, that complicates the evaluation significantly, reducing the strength of the natural experiment.

Chapter 6

1. This is an important number because the final bill passed by 219 to 212, so 12 defectors would have prevented the bill from passing.

2. The models throughout this book, with the exception of appendix A, have been based on a unidimensional policy space, but in this case, a multidimensional characterization may be more useful.

Chapter 7

1. Poole and Rosenthal (1997) estimate scores for presidents, but the estimates are weaker because most bills don't get to the presidential signature or veto stage. Moreover, Speakers rarely cast votes. Therefore, the 81st Congress is the best comparison for them.
2. "Entitlement spending" refers to spending on programs such as Social Security and Medicare. Money is automatically appropriated for these programs unless Congress alters the programs because eligible citizens cannot be blocked from payment.
3. This is not merely retrospective analysis. An earlier version of this model, claiming instability, was presented in April 2014 at the Midwest Political Science Association conference.
4. By 2013, that demand evolved into a repeal, defunding, or delay of the PPACA.
5. Unsurprisingly, the supercommittee failed because Democrats on the committee insisted that at least some of the deficit reduction come from revenue increases, while Republicans insisted that the package consist entirely of spending cuts. Sequestration then went into effect.
6. As an interesting postscript, Huelskamp was arguably the only incumbent in modern times to be successfully "primaried" as an extremist. In 2016, party establishment figures, fed up with Huelskamp's extremism, backed an establishment figure against him in a primary—Roger Marshall. Marshall defeated Huelskamp. The story, however, has not had the same disproportionate impact on incumbents' mentality as the story of Dick Lugar.
7. Additionally, the somewhat more moderate Representative Walter Jones (R-NC) was stripped of his committee assignment, which was an extreme measure, but less surprising given his relative moderation.
8. This points, once again, to the intrinsic ambiguity of a no vote, which can make it substantively different from a yes vote (Buchler 2017).

Appendix A

1. The unified model in chapter 4 assumed that voters presumed that challengers would vote in a manner consistent with their party medians based on party reputation, but once we open the model up to multidimensionality, this type of strategy must be considered.

Appendix B

1. Erikson and Romero (1990) consider the possibility that the partisanship dimension might not be orthogonal to the policy dimension, but it operates as a separate dimension.
2. In that respect, partisan voters will more closely resemble those in a model from Buchler (2011b), in which voters' decisions are fixed by party, so incumbents position themselves strategically based on threats of challenger entry.

REFERENCES

Abramowitz, Alan I. 2008. "Don't Blame Primary Voters for Polarization." *Forum* 5: Article 4.

Abramowitz, Alan I., Brad Alexander, and Matthew Gunning. 2006. "Incumbency, Redistricting, and the Decline of Competition in U.S. House Elections." *Journal of Politics* 68: 75–88.

Adams, James. 1999. "Policy Divergence in Multicandidate Probabilistic Spatial Voting." *Public Choice* 99: 259–74.

Adams, James, Samuel Merrill III, and Bernard Grofman. 2005. *A Unified Theory of Party Competition: A Cross-National Analysis Integrating Spatial and Behavioral Factors*. Cambridge: Cambridge University Press.

Aldrich, John H. 1983a. "A Spatial Model with Party Activists: Implications for Electoral Dynamics." *Public Choice* 41: 63–100.

Aldrich, John H. 1983b. "A Downsian Spatial Model with Party Activism." *American Political Science Review* 77: 974–990.

Aldrich, John H. 1995. *Why Parties?* Chicago: University of Chicago Press.

Aldrich, John H. and David W. Rohde. 2000. "The Consequences of Party Organization in the House: The Role of the Majority and Minority Parties in Conditional Party Government." In *Polarized Politics: Congress and the President in a Partisan Era*, ed. Jon R. Bond and Richard Fleisher. Washington, DC: Congressional Quarterly Press. 31–72.

Ansolabehere, Stephen and Philip Edward Jones. 2010. "Constituents' Responses to Congressional Roll Call Voting." *American Journal of Political Science* 54: 583–597.

Ansolabehere, Stephen and James M. Snyder Jr. 2000. "Valence Politics and Equilibrium in Spatial Election Models." *Public Choice* 103: 327–36.

Ansolabehere, Stephen, James M. Snyder Jr., and Charles Stewart III. 2001. "Candidate Positioning in U.S. House Elections." *American Journal of Political Science* 45: 136–159.

Austen-Smith, David. 1984. "Two-Party Competition with Many Constituencies." *Mathematical Social Sciences* 7: 177–198.

Austen-Smith, David. 1986. "Legislative Coalitions and Electoral Equilibrium." *Public Choice* 50: 185–210.

Bafumi, Joseph and Michael C. Herron. 2010. "Leapfrog Representation and Extremism: A Study of American Voters and Their Members in Congress." *American Political Science Review* 104: 519–542.

Bender, Bruce and John R. Lott. 1996. "Legislator Voting and Shirking: A Critical Review of the Literature." *Public Choice* 87: 67–100.

Bertelli, Anthony M. and Jeffrey B. Wenger. 2009. "Demanding Information: Think Tanks and the US Congress." *British Journal of Political Science* 39: 225–242.

Black, Duncan. 1958. *The Theory of Committees and Elections*. Cambridge: Cambridge University Press.

Boatright, Robert C. 2013. *Getting Primaried: The Changing Politics of Congressional Primary Challenges*. Ann Arbor: University of Michigan Press.

Bovitz, Gregory L. and Jamie L. Carson. 2006. "Position-Taking and Electoral Accountability in the US House of Representatives." *Political Research Quarterly* 59: 297–312.

Brady, David W., Hahrie Han, and Jeremy C. Pope. 2007. "Primary Elections and Candidate Ideology: Out of Step with the Primary Electorate?" *Legislative Studies Quarterly* 32: 79–106.

Brunell, Thomas L. 2006. "Rethinking Redistricting: How Drawing Uncompetitive Districts Eliminates Gerrymanders, Enhances Representation, and Improves Attitudes toward Congress." *PS: Political Science & Politics* 40: 77–85.

Brunell, Thomas L. 2008. *Redistricting and Representation: Why Competitive Elections Are Bad for America*. New York: Routledge.

Buchler, Justin. 2005. "Competition, Representation and Redistricting: The Case against Competitive Congressional Districts." *Journal of Theoretical Politics* 17: 431–463.

Buchler, Justin. 2009. "Redistricting Reform Will Not Solve California's Budget Crisis." *California Journal of Politics & Policy* 1: article 3.

Buchler, Justin. 2010a. "The Redistricting Process Should Be Nonpartisan: Con." In *Debating Reform*, ed. Richard Ellis and Michael Nelson. Washington, DC: Congressional Quarterly Press. 161–171.

Buchler, Justin. 2010b. "The Inevitability of Gerrymandering: Winners and Losers under Alternative Approaches to Redistricting." *Duke Journal of Constitutional Law and Public Policy* 5: 17–36.

Buchler, Justin. 2011a. "The Proximity Paradox: The Legislative Agenda and the Electoral Success of Ideological Extremists." *Public Choice* 148: 1–19.

Buchler, Justin. 2011b. *Hiring and Firing Public Officials: Rethinking the Purpose of Elections*. New York: Oxford University Press.

Buchler, Justin. 2017. "Expressive Voting and Legislative Gridlock: The Changing Meaning of a No Vote." Paper prepared for delivery at the 2017 Annual Meeting of the Midwest Political Science Association.

Butler, David and Bruce E. Cain. 1992. *Congressional Redistricting: Comparative and Theoretical Perspectives*. New York: Macmillan.

Campbell, Angus, Phillip E. Converse, Warren E. Miller, and Donald E. Stokes. 1960. *The American Voter*. New York: John Wiley & Sons.

Canes-Wrone, Brandice, David W. Brady, and John F. Cogan. 2002. "Out of Step, out of Office: Electoral Accountability and House Members' Voting." *American Political Science Review* 96: 127–140.

Carson, Jamie L. 2005. "Strategy, Selection, and Candidate Competition in U.S. House and Senate Elections." *Journal of Politics* 67: 1–28.

Carson, Jamie L., Gregory Koger, and Matthew J. Lebo. 2010. "The Electoral Costs of Party Loyalty in Congress." *American Journal of Political Science* 54: 598–616.

Converse, Phillip E. 1964. "The Nature of Belief Systems in Mass Publics." In *Ideology and Discontent*, ed. David Apter. New York: Free Press. 206–261.

Cox, Gary W. and Mathew D. McCubbins. 1993. *Legislative Leviathan: Party Government in the House*. Berkeley: University of California Press.

Cox, Gary W. and Mathew D. McCubbins. 1994. "Bonding, Structure, and the Stability of Political Parties: Party Government in the House." *Legislative Studies Quarterly* 19: 215–231.

Cox, Gary W. and Mathew D. McCubbins. 1997. "Toward a Theory of Legislative Rules Changes: Assessing Schickler and Rich's Evidence." *American Journal of Political Science* 41: 1376–1386.

Cox, Gary W. and Mathew D. McCubbins. 2002. "Agenda Power in the U.S. House of Representatives, 1877–1986." In *Party, Process, and the Political Change in Congress: New Perspectives on the History of Congress*, ed. David W. Brady and Mathew D. McCubbins. Stanford: Stanford University Press. 107–145.

Cox, Gary W. and Mathew D. McCubbins. 2005. *Setting the Agenda: Responsible Party Government in the U.S. House of Representatives*. Cambridge: Cambridge University Press.

Cukierman, Alex. 1991. "Asymmetric Information and the Electoral Momentum of Public Opinion Polls." *Public Choice* 70: 181–213.

DeBacker, Jason. 2012. "Political Parties and Political Shirking." *Public Choice* 150: 651–670.

Downs, Anthony. 1957. *An Economic Theory of Democracy*. New York: HarperCollins.

Enelow, James M. and Melvin J. Hinich. 1982. "Nonspatial Candidate Characteristics and Electoral Competition." *Journal of Politics* 33: 115–130.

Erikson, Robert S. and David W. Romero. 1990. "Candidate Equilibrium and the Behavioral Model of the Vote." *American Political Science Review* 84: 1103–1126.

Erikson, Robert S. and Gerald C. Wright. 2000. "Representation of Constituency Ideology in Congress." In *Change and Continuity in House Elections*, ed. David W. Brady, John F. Cogan, and Morris P. Fiorina. Stanford: Stanford University Press. 149–177.

Fenno, Richard F., Jr. 1973. *Congressmen in Committees*. Boston: Little, Brown.

Fiorina, Morris P. 1973. "Electoral Margins, Constituency Influence and Policy Moderation: A Critical Assessment." *American Politics Quarterly* 1: 479–498.

Fowler, Linda L. and Brian Frederking. 2000. "Representation, Careerism, and Term Limits: A Simulation." In *Congress on Display, Congress at Work*, ed. William T. Bianco. Ann Arbor: University of Michigan Press. 189–215.

Gilmour, John B. 1995. *Strategic Disagreement: Stalemate in American Politics*. Pittsburgh: University of Pittsburgh Press.

Green, Donald P., Bradley Palmquist, and Eric Schickler. 2002. *Partisan Hearts and Minds: Political Parties and the Social Identities of Voters*. New Haven: Yale University Press.

Griffin, John. 2006. "Electoral Competition and Democratic Responsiveness: A Defense of the Marginality Hypothesis." *Journal of Politics* 68: 911–921.

Grofman, Bernard, William Koetzle, and Anthony J. McGann. 2002. "Congressional Leadership 1965–96: A New Look at the Extremism versus Centrality Debate." *Legislative Studies Quarterly* 27: 87–105.

Groseclose, Tim. 2001. "A Model of Candidate Location When One Candidate Has a Valence Advantage." *American Journal of Political Science* 45(4): 862–886.

Groseclose, Tim and Jeffrey Milyo. 2010. "Sincere versus Sophisticated Voting in Congress." *Journal of Politics* 72: 60–73.

Harden, Jeffrey J. and Thomas M. Carsey. 2012. "Balancing Constituency Representation and Party Responsiveness in the US Senate: The Conditioning Effect of State Ideological Heterogeneity." *Public Choice* 150: 137–154.

Hirano, Shigeo, James M. Snyder Jr., and Stephen Ansolabehere. 2010. "Primary Elections and Partisan Polarization in the US Congress." *Quarterly Journal of Political Science* 5: 169–191.

Jacobson, Gary C. 2000. "Party Polarization in National Politics: The Electoral Connection." In *Polarized Politics: Congress and the President in a Partisan Era*, ed. John R. Bond and Richard Fleisher. Washington, DC: Congressional Quarterly Press.

Jenkins, Jeffrey A., Michael H. Crespin, and Jamie L. Carson. 2005. "Parties as Procedural Coalitions in Congress: An Examination of Differing Career Tracks." *Legislative Studies Quarterly* 30: 365–389.

Jenkins, Jeffrey A. and Timothy P. Nokken. 2008. "Partisanship, the Electoral Connection, and Lame Duck Sessions of Congress: 1877–2006." *Journal of Politics* 70: 450–465.

Jenkins, Jeffrey A. and Charles Stewart III. 2012. *Fighting for the Speakership: The House and the Rise of Party Government*. Princeton: Princeton University Press.

Jessee, Stephen A. 2012. *Ideology and Spatial Voting in American Elections*. New York: Cambridge University Press.

Kartik, Navin and R. Preston McAfee. 2007. "Signaling Character in Electoral Competition." *American Economic Review* 97: 852–870.

Krehbiel, Keith. 1998. *Pivotal Politics: A Theory of U.S. Lawmaking*. Chicago: University of Chicago Press.

Lacy, Dean and Phillip Paolino. 1998. "Downsian Voting and the Separation of Powers." *American Journal of Political Science* 42: 1180–99.
Ladewig, Jeffrey W. 2005. "Conditional Party Government and the Homogeneity of Constituent Interests." *Journal of Politics* 67: 1006–1029.
Ladewig, Jeffrey W. 2010. "Ideological Polarization and the Vanishing Marginals: Retrospective Roll Call Voting in the U.S. Congress." *Journal of Politics* 72: 499–512.
Lebo, Matthew J., Adam J. McGlynn, and Gregory Koger. 2007. "Strategic Party Government: Party Influence in Congress, 1789–2000." *American Journal of Political Science* 51 (3): 464–481).
Lee, Frances. 2009. *Beyond Ideology: Politics, Principles and Partisanship in the U.S. Senate*. Chicago: University of Chicago Press.
Londregan, John and Thomas Romer. 1993. "Polarization, Incumbency, and the Personal Vote." In *Political Economy: Institutions, Competition and Representation*, ed. William A. Barnett, Melvin J. Hinich, and Norman J. Schofield. New York: Cambridge University Press. 355–377.
Lott, John R. 1987. "Political Cheating." *Public Choice* 52: 169–186.
Lott, John R. and Michael Davis. 1992. "A Critical Review and an Extension of the Political Shirking Literature." *Public Choice* 61: 75–96.
Lott, John R. and W. Robert Reed. 1989. "Shirking and Sorting in a Political Market with Finite-Lived Politicians." *Public Choice* 61: 75–96.
Mann, Thomas E. 1978. *Unsafe at Any Margin: Interpreting Congressional Elections*. Washington, DC: American Enterprise Institute.
Mann, Thomas E. and Norman Ornstein. 2012. *It's Even Worse Than It Looks: How the American Constitutional System Collided with the New Politics of Extremism*. New York: Basic Books.
Masket, Seth. 2009. *No Middle Ground: How Informal Party Organizations Control Nominations and Polarize Legislatures*. Ann Arbor: University of Michigan Press.
Mayhew, David. 1974. *Congress: The Electoral Connection*. New Haven: Yale University Press.
Mayhew, David. 1991. *Divided We Govern*. New Haven: Yale University Press.
McCarty, Nolan, Keith T. Poole and Howard Rosenthal. 2009. "Does Gerrymandering Cause Polarization?" *American Journal of Political Science* 53: 666–680.
McKelvey, Richard D. 1976. "Intransitivities in Multidimensional Voting Models and Some Implications for Agenda Control." *Journal of Economic Theory* 12: 472–482.
Noel, Hans. 2014. *Political Ideologies and Political Parties in America*. Cambridge: Cambridge University Press.
Nyhan, Brendan, Eric McGhee, John Sides, Seth Masket, and Steven Greene. 2012. "One Vote out of Step? The Effects of Salient Roll Call Votes in the 2010 Election." *American Politics Research* 20: 1–36.
Olson, Mancur. 1966. *The Logic of Collective Action*. Cambridge: Harvard University Press.
Page, Benjamin I. 1978. *Choices and Echoes in Presidential Elections*. Chicago: University of Chicago Press.
Parker, Glenn R. 2008. *Capitol Investments: The Marketability of Political Skills*. Ann Arbor: University of Michigan Press.
Patty, John W. 2008. "Equilibrium Party Government." *American Journal of Political Science* 52: 636–655.
Pearson, Kathryn. 2015. *Party Discipline in the U.S. House of Representatives*. Ann Arbor: University of Michigan Press.
Poole, Keith T. and Howard Rosenthal. 1997. *Congress: A Political-Economic History of Roll Call Voting*. New York: Oxford University Press.
Richman, Jesse. 2011. "Parties, Pivots and Policy: The Status Quo Test." *American Political Science Review* 105: 151–165.
Roemer, John E. 2001. *Political Competition: Theory and Applications*. Cambridge: Harvard University Press.
Rohde, David W. 1991. *Parties and Leaders in the Postreform House*. Chicago: University of Chicago Press.

Rothenberg, L. and M. Sanders. 2000. "Severing the Electoral Connection: Shirking in the Contemporary Congress." *American Journal of Political Science* 44: 310–319.

Schofield, Norman. 1978. "Instability of Simple Dynamic Games." *Review of Economic Studies* 45: 575–594.

Schofield, Norman. 2004. "Equilibrium in the Spatial 'Valence' Model of Politics." *Journal of Theoretical Politics* 16(4): 447–481.

Schofield, Norman and Itai Sened. 2005a. "Multiparty Competition in Israel, 1988– 1996." *British Journal of Political Science* 35(4): 635–663.

Schofield, Norman and Itai Sened. 2005b. "Modeling the Interaction of Parties, Activists and Voters: Why Is the Political Center So Empty?" *European Journal of Political Research* 44(3): 355–390.

Shepsle, Kenneth A. 1972. "The Strategy of Ambiguity: Uncertainty and Electoral Competition." *American Political Science Review* 66: 555–68.

Shepsle, Kenneth A. and Barry Weingast. 1981. "Structure Induced Equilibrium and Legislative Choice." *Public Choice* 37: 503–519.

Sin, Gisela. 2014. *Separation of Powers and Legislative Organization: The President, the Senate, and Political Parties in the Making of House Rules.* Cambridge: Cambridge University Press.

Sinclair, Barbara. 2011. *Unorthodox Lawmaking: New Legislative Processes in the U.S. Congress.* 4th edition. Washington, DC: Congressional Quarterly Press.

Skocpol, Theda and Vanessa Williamson. 2012. *The Tea Party and the Remaking of Republican Conservatism.* New York: Oxford University Press.

Smith, Richard A. 1995. "Interest Group Influence in the U.S. Congress." *Legislative Studies Quarterly* 20(1): 89–139.

Smith, Steven S. 2007. *Party Influence in Congress.* Cambridge: Cambridge University Press.

Sniderman, Paul M. and Edward H. Stiglitz. 2012. *The Reputational Premium: A Theory of Party Identification and Policy Reasoning.* Princeton: Princeton University Press.

Snyder, James M. and Tim Groseclose. 2000. "Estimating Party Influence on Roll Call Voting." *American Journal of Political Science* 44(2): 193–211.

Snyder, James M. and Michael Ting. 2002. "An Informational Rationale for Political Parties." *American Journal of Political Science* 46: 90–110.

Stokes, Donald. 1963. "Spatial Models of Party Competition." *American Political Science Review* 57: 368–377.

Sullivan, John L. and Eric M. Uslaner. 1978. "Congressional Behavior and Electoral Marginality." *American Journal of Political Science* 22: 536–553.

Theriault, Sean M. 2008. *Party Polarization in Congress.* Cambridge: Cambridge University Press.

Tomz, Michael and Robert P. Van Houweling. 2009. "The Electoral Implications of Candidate Ambiguity." *American Political Science Review* 103: 83–98.

Van Houweling, Robert. 2012. "Parties as Enablers." Unpublished manuscript.

Volden, Craig and Elizabeth Bergman. 2006. "How Strong Should Our Party Be? Party Member Preferences over Party Cohesion." *Legislative Studies Quarterly* 31: 71–104.

Wittman, Donald. 1973. "Parties as Utility Maximizers." *American Political Science Review* 67: 490–498.

Wittman, Donald. 1977. "Candidates with Policy Preferences: A Dynamic Model." *Journal of Economic Theory* 14: 180–189.

Wittman, Donald. 1983. "Candidate Motivation: A Synthesis of Alternative Theories." *American Political Science Review* 77: 142–157.

Wright, Gerald C., Robert S. Erikson, and John P. McIver. 1985. "Measuring State Partisanship and Ideology with Survey Data." *Journal of Politics* 47: 469–489.

INDEX

The manufacturer's authorised representative in the EU for product
safety is Oxford University Press España S.A. of El Parque Empresarial
San Fernando de Henares, Avenida de Castilla, 2 - 28830 Madrid
(www.oup.es/en or product.safety@oup.com). OUP España S.A. also acts
as importer into Spain of products made by the manufacturer.
Printed and bound by CPI Group (UK) Ltd, Croydon, CR0 4YY
14/09/2026
02216306-0017